Refrigerant Transition and Recovery Certification: Program Manual for HVACR Technicians, 5E

Delta-T Solutions, Inc.

DELMAR

THOMSON LEARNING ™

Australia Canada Mexico Singapore Spain United Kingdom United States

DELMAR

THOMSON LEARNING

Refrigerant Transition and Recovery Certification:
Program Manual for HVACR Technicians, 5E
Delta-T Solutions, Inc.

Business Unit Director: Alar Elken	**Executive Marketing Manager:** Maura Theriault	**Channel Manager:** Mona Caron
Executive Editor: Sandy Clark	**Executive Production Manager:** Mary Ellen Black	**Marketing Coordinator:** Brian McGrath
Acquisitions Editor: Sanjeev Rao	**Production Manager:** Larry Main	**Cover Design:** Cammi Noah
Team Assistant Matthew Seeley	**Production Editor** Tom Stover	

NOTICE TO THE READER

REFRIGERANT TRANSITION AND RECOVERY CERTIFICATION PROGRAM MANUAL
FOR HVACR TECHNICIANS

This manual contains educational materials that exceed Environmental Protection Agency (EPA) certification testing requirements. We believe that this value-added information is important to technicians in the transition to alternative refrigerants and oils and the solution to stratospheric ozone depletion and global warming.

Persons successfully completing a core of questions on stratospheric ozone protection (25 each) and questions from one or more of the following types at 70 percent success, will be certified in those areas. Persons certified in all three area types will be universally certified. Staying abreast of future changes in regulations and technology is the responsibility of the technician.

Type I Certification

  Small Appliance — Manufactured, charged and hermetically sealed with five (5) pounds or less of refrigerant. Includes refrigerators, freezers, room air conditioners, package terminal heat pumps, dehumidifiers, under-the-counter ice makers, vending machines, and drinking water coolers.

Type II Certification

  High Pressure Appliance — Uses refrigerant with a boiling point between -50°C (-58°F) and 10°C (50°F) at atmospheric pressure. Includes 12, 22, 114, 500, and 502 refrigerants.

  Very High Pressure Appliance — Uses refrigerants with a boiling point below -50°C (-58°F) at atmospheric pressure. Includes 13 and 503 refrigerants.

Type III Certification

  Low Pressure Appliance — Uses refrigerant with a boiling point above 10°C (50°F) at atmospheric pressure. Includes 11, 113, and 123 refrigerants.

Universal Certification

  Certified in all the above: Type I, II, and III.

For regulatory updates and additional information:—Online Information From EPA:
  Definitions:—http://www.epa.gov/spdpublc/defns.html#strat
  Common Questions:
      —http://www.al.noaa.gov/WWWHD/pubdocs/Assessment94/common-questions.html
  Complying with The Section 608 Refrigerant Recycling Rule:
      —http://www.epa.gov/ozone/title6/608/608fact.html
  Ozone-Depleting Substances:—http://www.epa.gov/spdpublc/ods.html

Preface

This edition of the *Refrigerant Transition and Recovery Certification Program Manual* was written by two current faculty members and one Emeriti faculty member of the HVACR Department at Ferris State University, Big Rapids, Michigan. A practical application of refrigeration and air conditioning system technology is emphasized in this edition with many illustrations for ease in understanding important concepts. It is produced and designed to educate beginning technicians as well as seasoned veterans in the transition to environmentally safer refrigerants and oils, helping technicians to become certified in refrigerant handling as mandated by 40 CFR, Part 82, Subpart F of the Clean Air Act amendments. Included in this manual are the appropriate air conditioning and refrigeration systems which can prepare participants to be certified in small appliance, high and very high pressure systems, and low pressure systems.

At time of printing, the information on alternate and interim refrigerants and oils was the current technology. This manual will be revised on a reasonable schedule to reflect industry and EPA regulatory changes in the examination.

This project was conducted in cooperation with numerous manufacturers, most of which are listed with acknowledgments in this manual. Their assistance made the solutions portion of this manual possible.

AUTHORS
John Tomczyk, Professor, FSU
Joe Nott, Associate Professor, FSU
Dick Shaw, Professor Emeritus, FSU

Acknowledgments

Air Conditioning and Refrigeration Institute (ARI)
The Air Conditioning, Heating and Refrigeration News
American Society of Heating, Refrigeration and Air Conditioning Engineers, Inc. (ASHRAE)
Blissfield Manufacturing Company
Bohn Heat Transfer Company
BVA Oils
Carlyle
Carrier
Castrol
Clean Air Act Amendments of 1990 (P.L. 101-549)
Copeland
Danfoss Inc.
The Delfield Company
DuPont Chemicals
Frigidaire Company
General Motors
Honeywell
Merit Mechanical Systems, Inc.
Mobil
National Refrigerants
Newsweek
Refrigeration Research, Inc.
Robinair
Scientific American
Tecumseh Corporation
Thermal Engineering Company
TIF Corporation
Time
Trane
U.S. Department of Transportation
U.S. Environmental Protection Agency

Table of Contents

SECTIONS

SECTION ONE
Refrigeration and Air Conditioning Systems Fundamentals

Condensing Pressure	Evaporating Pressure
High side pressure	Low side pressure
Head pressure	Suction pressure
Discharge pressure	Back pressure

Vapor Compression Refrigeration System

Refrigeration is defined as that branch of science which deals with the process of reducing and maintaining the temperature of a space or materials below the temperature of the surroundings. To accomplish this, heat must be removed from the refrigerated body and transferred to another body.

The typical vapor compression refrigeration system shown in Figure 1-1 can be divided into two pressures: condensing (high side) and evaporating (low side). These pressures are divided or separated in the system by the compressor discharge valve and the metering device. Listed at the top of the next column are field service terms often used to describe these pressures:

Condensing Pressure

The condensing pressure is the pressure at which the refrigerant changes state from a vapor to a liquid. This phase change is referred to as *condensation*. This pressure can be read directly from a pressure gauge connected anywhere between the compressor discharge valve and the entrance to the metering device, assuming there is negligible pressure drop. In reality, line and valve friction and the weight of the liquid itself cause pressure drops from the compressor discharge to the metering device. If the true condensing pressure is needed, the technician must measure the pressure as close to the condenser as possible to avoid these pressure drops. This pressure is usually measured on smaller systems near the compressor valves, Figure 1-2. On small systems, it is not critical where a technician places the pressure gauge (as long as it is on the high side of the system), because pressure drops are negligible. The pressure gauge reads the same no matter where it is on the high side of the system if line and valve losses are negligible.

Evaporating Pressure

The evaporating pressure is the pressure at which the refrigerant changes state from a liquid to a vapor. This phase change is referred to as *evaporation* or *vaporizing*. A pressure gauge placed anywhere between the metering device outlet and the compressor (including compressor crankcase) will read the evaporating pressure. Again, negligible pressure drops are assumed. In reality, there will be line and valve pressure drops as the refrigerant travels through the evaporator and suction line. The technician must measure the pressure as close to the

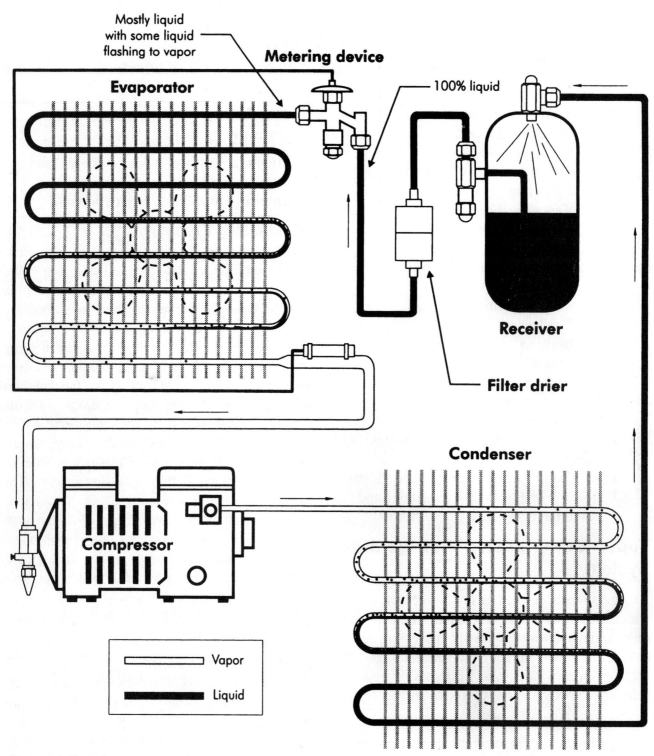

Figure 1-1. Typical compression refrigeration system

evaporator as possible to get a true evaporating pressure. On small systems where pressure drops are negligible, this pressure is usually measured near the compressor (see Figure 1-2). Gauge placement on small systems is usually not critical as long as it is placed on the low side of the refrigeration system, because the refrigerant vapor pressure acts equally in all directions. If line and valve pressure drops become substantial, gauge placement becomes

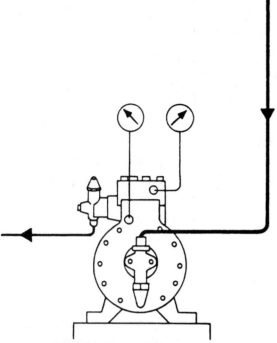

Figure 1-2. Semi-hermetic compressor showing pressure access valves (Courtesy, Danfoss Automatic Controls, Division of Danfoss, Inc.)

critical. In larger more sophisticated systems, gauge placement is more critical because of associated line and valve pressure losses. If the system has significant line and valve pressure losses, the technician must place the gauge as close as possible to the component that requires a pressure reading.

Refrigerant States and Conditions

Modern refrigerants exist either in the vapor or liquid state. Refrigerants have such low freezing points that they are rarely in the frozen or solid state. Refrigerants can co-exist as vapor and liquid as long as conditions are right. Both the evaporator and condenser house liquid and vapor refrigerant simultaneously if the system is operating properly. Refrigerant liquid and vapor can exist in both the high or low pressure sides of the refrigeration system.

Along with refrigerant pressures and states are refrigerant conditions. Refrigerant conditions can be *saturated, superheated,* or *subcooled.*

Saturation

Saturation is usually defined as a temperature. The saturation temperature is the temperature at which a fluid changes from liquid to vapor or vapor to liquid. At saturation temperature, liquid and vapor are called saturated liquid and saturated vapor, respectively. Saturation occurs in both the evaporator and condenser. At saturation, the liquid experiences its maximum temperature for the pressure, and the vapor experiences its minimum temperature. However, both liquid and vapor are at the same temperature for the given pressure when saturation occurs. Saturation temperatures vary with different refrigerants and pressures. All refrigerants have different vapor pressures. It is vapor pressure that is measured with a gauge.

Vapor Pressure

Vapor pressure is the pressure exerted on a saturated liquid. Any time saturated liquid and vapor are together (as in the condenser and evaporator), vapor pressure is generated. Vapor pressure acts equally in all directions and affects the entire low or high side of a refrigeration system.

As pressure increases, saturation temperature increases; as pressure decreases, saturation temperature decreases. Only at saturation are there pressure/temperature relationships for refrigerants. Table 1-1 shows the pressure/temperature relationship at saturation for refrigerant 134a (R-134a). If one attempts to raise the temperature of a saturated liquid above its saturation temperature, vaporization of the liquid will occur. If one attempts to lower the temperature of a saturated vapor below its saturation temperature, condensation will occur. Both vaporization and condensation occur in the evaporator and condenser, respectively.

The heat energy that causes a liquid refrigerant to change to a vapor at a constant saturation temperature for a given pressure is referred to as *latent heat.* Latent heat is the heat energy that causes a substance to change state without changing the temperature of the substance. Vaporization and condensation are examples of a latent heat process.

Temperature (°F)	Pressure (psig)	Temperature (°F)	Pressure (psig)
-10	1.8	25	21.7
-9	2.2	26	22.4
-8	2.6	27	23.2
-7	3.0	28	24.0
-6	3.5	29	24.8
-5	3.9	30	25.6
-4	4.4	31	26.4
-3	4.8	32	27.3
-2	5.3	33	28.1
-1	5.8	34	29.0
0	6.2	35	29.9
1	6.7	40	34.5
2	7.2	45	39.5
3	7.8	50	44.9
4	8.3	55	50.7
5	8.8	60	56.9
6	9.3	65	63.5
7	9.9	70	70.7
8	10.5	75	78.3
9	11.0	80	86.4
10	11.6	85	95.0
11	12.2	90	104.2
12	12.8	95	113.9
13	13.4	100	124.3
14	14.0	105	135.2
15	14.7	110	146.8
16	15.3	115	159.0
17	16.0	120	171.9
18	16.7	125	185.5
19	17.3	130	199.8
20	18.0	135	214.8
21	18.7		
22	19.4		
23	20.2		
24	20.9		

Table 1-1. R-134a saturated vapor/liquid pressure/ temperature chart

Superheat

Superheat always refers to a vapor. A superheated vapor is any vapor that is above its saturation temperature for a given pressure. In order for vapor to be superheated, it must have reached its 100% saturated vapor point. In other words, all of the liquid must be vaporized for superheating to occur; the vapor must be removed from contact with the vaporizing liquid. Once all the liquid has been vaporized at its saturation temperature, any addition of heat causes the 100% saturated vapor to start superheating. This addition of heat causes the vapor to increase in temperature and gain *sensible heat*. Sensible heat is the heat energy that causes a change in the temperature of a substance. The heat energy that superheats vapor and increases its temperature is sensible heat energy. Superheating is a sensible heat process. Superheated vapor occurs in the evaporator, suction line, and compressor.

Subcooling

Subcooling always refers to a liquid at a temperature below its saturation temperature for a given pressure. Once all of the vapor changes to 100% saturated liquid, further removal of heat will cause the 100% liquid to drop in temperature or lose sensible heat. Subcooled liquid results. Subcooling can occur in both the condenser and liquid line and is a sensible heat process.

A thorough understanding of pressures, states, and conditions of the basic refrigeration system enables the service technician to be a good systematic troubleshooter. It is not until then that a service technician should even attempt systematic troubleshooting.

Basic Refrigeration System

Figure 1-3 illustrates a basic refrigeration system. The basic components of this system are the compressor, discharge line, condenser, receiver, liquid line, metering device, evaporator, and suction line. Mastering the function of each individual component can assist the refrigeration technician with analytical troubleshooting skills, saving time and money for both technician and customer.

Compressor

One of the main functions of the compressor is to circulate refrigerant. Without the compressor as a refrigerant pump, refrigerant could not reach other system components to perform its heat transfer functions. The compressor also separates the high pressure from the low pressure side of the refrigeration system. A difference in pressure is mandatory for fluid (gas or liquid) flow, and there can be no refrigerant flow without this pressure separation.

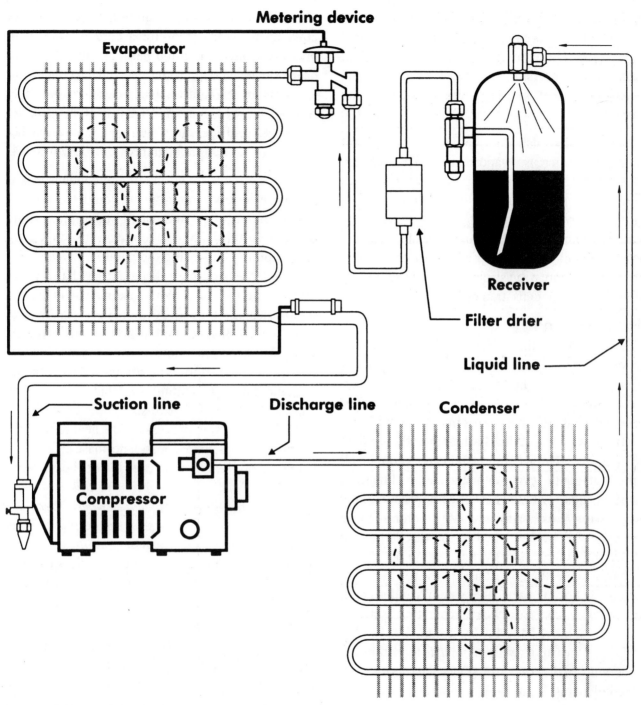

Figure 1-3. Basic refrigeration system

Another function of the compressor is to elevate or raise the temperature of the refrigerant vapor above the ambient (surrounding) temperature. This is accomplished by adding work, or heat of compression, to the refrigerant vapor during the compression cycle. The pressure of the refrigerant is raised, as well as its temperature. By elevating the refrigerant temperature above the ambient temperature, heat absorbed in the evaporator and suction line, and any heat of compression generated in the compression stroke can be rejected to this lower temperature ambient. Most of the heat is rejected in the discharge line and the condenser.

Remember, heat flows from hot to cold, and there must be a temperature difference for any heat transfer to take place. The temperature rise of the refrigerant during the compression stroke is a measure of the increased internal kinetic energy added by the compressor.

The compressor also compresses the refrigerant vapors, which increases vapor density. This increase in density helps pack the refrigerant gas molecules together, which helps in the condensation or liquification of the refrigerant gas molecules in the condenser once the right amount of heat is rejected to the ambient. The compression of the vapors during the compression stroke is actually preparing the vapors for condensation or liquification.

Discharge Line

One function of the discharge line is to carry the high pressure superheated vapor from the compressor discharge valve to the entrance of the condenser. The discharge line also acts as a desuperheater, cooling the superheated vapors that the compressor has compressed and giving that heat up to the ambient surroundings. These compressed vapors contain all of the heat that the evaporator and suction line have absorbed, along with the heat of compression of the compression stroke. Any generated motor winding heat may also be contained in the discharge line refrigerant, which is why the beginning of the discharge line is the hottest part of the refrigeration system. On hot days when the system is under a high load and may have a dirty condenser, the discharge line can reach over 400°F. By desuperheating the refrigerant, the vapors will be cooled to the saturation temperature of the condenser. Once the vapors reach the condensing saturation temperature for that pressure, condensation of vapor to liquid will take place as more heat is lost.

Condenser

The first passes of the condenser desuperheat the discharge line gases. This prepares the high pressure superheated vapors coming from the discharge line for condensation, or the phase change from gas to liquid. Remember, these superheated gases must lose all of their superheat before reaching the condensing temperature for a certain condensing pressure. Once the initial passes of the condenser have rejected enough superheat and the condensing temperature or saturation temperature has been reached, these gases are referred to as 100% saturated vapor. The refrigerant is then said to have reached the 100% saturated vapor point (Point #2, Figure 1-4).

One of the main functions of the condenser is to condense the refrigerant vapor to liquid. Condensing is system dependent and usually takes place in the lower two-thirds of the condenser. Once the saturation or condensing temperature is reached in the condenser and the refrigerant gas has reached 100% saturated vapor, condensation can take place if more heat is removed. As more heat is taken away from the 100% saturated vapor, it will force the vapor to become a liquid or to condense. When condensing, the vapor will gradually phase change to liquid until 100% liquid is all that remains. This phase change, or change of state, is an example of a latent heat rejection process, as the heat removed is latent heat, not sensible heat. The phase change will happen at one temperature even though heat is being removed. *Note: An exception to this is a near-azeotropic blend of refrigerants where there is a temperature glide or range of temperatures when phase changing (see blend temperature glide in Section Two).* This one temperature is the saturation temperature corresponding to the saturation pressure in the condenser. As mentioned before, this pressure can be measured anywhere on the high side of the refrigeration system as long as line and valve pressure drops and losses are negligible.

The last function of the condenser is to subcool the liquid refrigerant. Subcooling is defined as any sensible heat taken away from 100% saturated liquid. Technically, subcooling is defined as the difference between the measured liquid temperature and the liquid saturation temperature at a given pressure. Once the saturated vapor in the condenser has phase changed to saturated liquid, the 100% saturated liquid point has been reached. If any more heat is

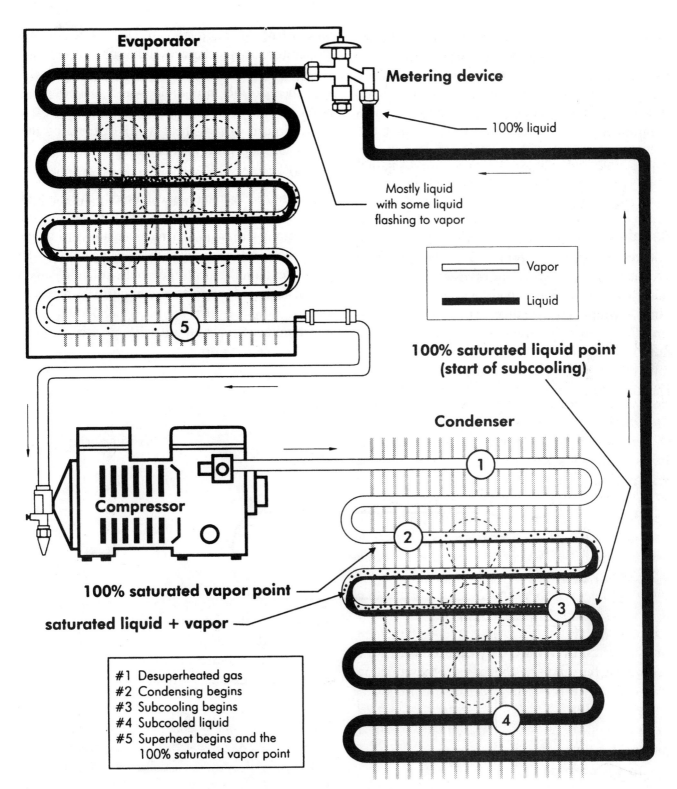

Evaporator

Metering device

100% liquid

Mostly liquid
with some liquid
flashing to vapor

5

Vapor

Liquid

**100% saturated liquid point
(start of subcooling)**

Condenser

1

2

100% saturated vapor point

saturated liquid + vapor

3

#1	Desuperheated gas
#2	Condensing begins
#3	Subcooling begins
#4	Subcooled liquid
#5	Superheat begins and the 100% saturated vapor point

Compressor

4

Figure 1-4. Basic refrigeration system showing 100% saturated vapor and liquid points

removed, the liquid will go through a sensible heat rejection process and lose temperature as it loses heat. The liquid that is cooler than the saturated liquid in the condenser is subcooled liquid. Subcooling is an important process, because it starts to lower the liquid temperature to the evaporator temperature. This will reduce flash loss in the evaporator so more of the vaporization of the liquid in the evaporator can be used for useful cooling of the product load.

Receiver

The receiver acts as a surge tank. Once the subcooled liquid exits the condenser, the receiver receives and stores the liquid. The liquid level in the receiver varies depending on whether the metering device is throttling opened or closed. Receivers are usually used on systems in which a thermostatic expansion valve (TXV or TEV) is used as the metering device. The subcooled liquid in the receiver may lose or gain subcooling depending on the surrounding temperature of the receiver. If the subcooled liquid is warmer than receiver surroundings, the liquid will reject heat to the surroundings and subcool even more. If the subcooled liquid is cooler than the receiver surroundings, heat will be gained by the liquid and subcooling will be lost.

A receiver bypass is often used to bypass liquid around the receiver and route it directly to the liquid line and filter drier. This bypass prevents subcooled liquid from sitting in the receiver and losing its subcooling. A thermostat with a sensing bulb on the condenser outlet controls the bypass solenoid valve by sensing liquid temperature coming to the receiver (Figure 1-5). If the liquid is subcooled to a predetermined temperature, it will bypass the receiver and go to the filter drier.

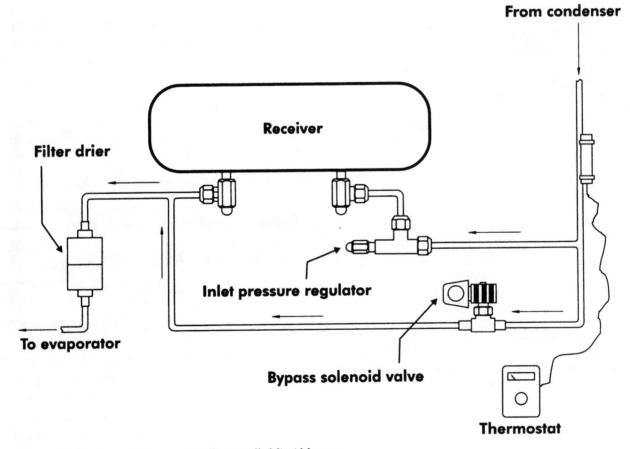

Figure 1-5. Receiver with thermostatically controlled liquid bypass

Figure 1-6. Liquid/suction line heat exchanger (Courtesy, Refrigeration Research, Inc.)

Liquid Line

The liquid line transports high pressure subcooled liquid to the metering device. In transport, the liquid may either lose or gain subcooling depending on the surrounding temperature. Liquid lines may be wrapped around suction lines to help them gain more subcooling (Figure 1-6). Liquid/suction line heat exchangers can be purchased and installed in existing systems to gain subcooling.

Metering Device

The metering device meters liquid refrigerant from the liquid line to the evaporator. There are several different styles and kinds of metering devices on the market with different functions. Some metering devices control evaporator superheat and pressure, and some even have pressure limiting devices to protect compressors at heavy loads.

The metering device is a restriction that separates the high pressure side from the low pressure side in a refrigeration system. The compressor and the metering device are the two components that separate pressures in a refrigeration system. The restriction in the metering device causes liquid refrigerant to flash to a lower temperature in the evaporator because of its lower pressure and temperatures.

Evaporator

The evaporator, like the condenser, acts as a heat exchanger. Heat gains from the product load and outside ambient travel through the sidewalls of the evaporator to vaporize any liquid refrigerant. The pressure drop through the metering device causes vaporization of some of the refrigerant and causes a lower saturation temperature in the evaporator. This temperature difference between the lower pressure refrigerant and the product load is the driving potential for heat transfer to take place.

The last pass of the evaporator coil acts as a superheater to ensure all liquid refrigerant has been vaporized. This protects the compressor from any liquid slopover, which may result in valve damage or diluted oil in the crankcase. The amount of superheat in the evaporator is usually controlled by a thermostatic expansion type of metering device.

Suction Line

The suction line transports low pressure superheated vapor from the evaporator to the compressor. There may be other components in the suction line such as suction accumulators, crankcase regulators, p-traps, filters, and screens. Liquid/suction line heat exchangers are often mounted in the suction line to transfer heat away from the liquid line (subcool) and into the suction line (Figure 1-7).

Another function of the suction line is to superheat the vapor as it approaches the compressor. Even though suction lines are usually insulated, sensible heat still penetrates the line and adds more superheat. This additional superheat decreases the density of the refrigerant vapor to prevent compressor overload, resulting in lower amp draws. This additional superheat also helps ensure that the compressor will see vapor only under low loading conditions. Many metering devices have a tendency to lose control of evaporator superheat at low loads. It is recommended that systems should have at least 20°F of total superheat at the compressor to prevent liquid slugging and /or flooding of the compressor at low loadings.

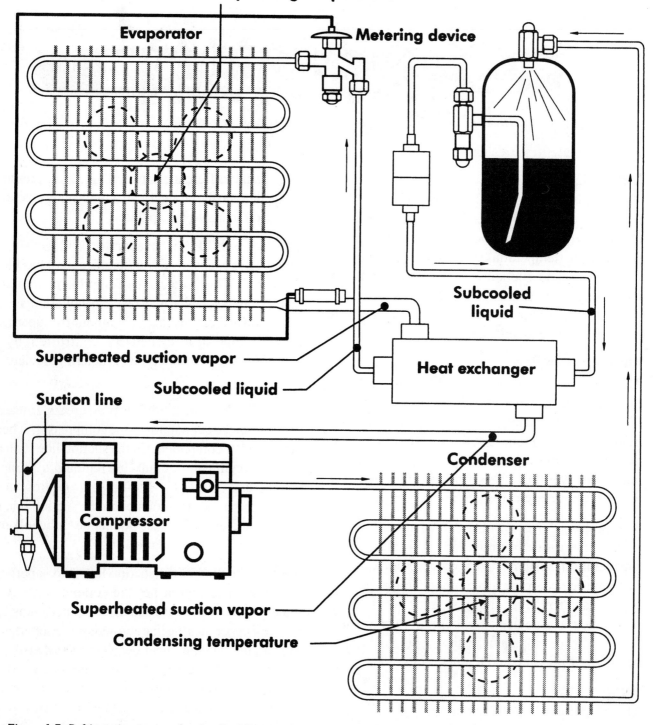

Figure 1-7. Refrigeration system showing liquid/suction line heat exchanger

Application of Pressures, States, and Conditions

Systematic troubleshooting requires mastering the function of all refrigeration system components. It is also important to be able to recognize the pressure, state, and condition of the working fluid (refrigerant) in the refrigeration system components. Figure 1-8 illustrates the basic refrigeration system. The legend lists refrigerant pressures, conditions, and states for the points shown in Figure 1-8. An explanation of the pressure, condition, and state of each point should clarify any system weaknesses.

Assume the following conditions for Figure 1-8:

Refrigerant = R-134a
Discharge (condensing) pressure = 124 psig (100°F)
Suction (evaporating) pressure = 6 psig (0°F)
Discharge temperature = 180°F
Condenser outlet temperature = 90°F
TXV inlet temperature = 80°F
Evaporator outlet temperature = 10°F
Compressor inlet temperature = 40°F

See Table 1-1 for pressure/temperature relationships.

Compressor Discharge (Point #1)

The refrigeration compressor is a vapor pump, not a liquid pump. The vapor leaving the compressor will be high pressure superheated vapor. The compressor is one of the two components in the system that separates the high pressure side from the low pressure side. The compressor discharge is high pressure, and the compressor suction is low pressure. The compressor discharge vapor receives its superheat from sensible heat coming from the evaporator, suction line, motor windings, friction, and internal heat of compression from the compression stroke. Since the vapor is superheated, no pressure/temperature relationship exists. Its temperature (180°F) is well above the saturation temperature of 100°F for the given saturation (condensing) pressure of 124 psig.

Condenser Inlet (Point #2)

As high pressure superheated refrigerant leaves the compressor, it instantly begins to lose superheat and cool in temperature. Its heat is usually given up to the surroundings. As mentioned before, this process is called desuperheating. Even though this refrigerant vapor is going through a desuperheating process, it is still superheated vapor. Pressure acts equally in all directions, so the vapor will be high pressure, or the same pressure as the compressor discharge, assuming that any line and valve pressure drops are ignored. Remember, the refrigerant is superheated and not saturated, so there is no pressure/temperature relationship. This high pressure superheated vapor is also above its saturation temperature of 100°F for the given discharge or condensing pressure of 124 psig. This point can be referred to as high pressure superheated vapor. This process of desuperheating will continue until the 100% saturated vapor point in the condenser is reached.

100% Saturated Vapor Point (Point #3)

Once all the superheat is rejected from the refrigerant gas, the saturation temperature of 100°F is finally reached for the condensing pressure of 124 psig (see Table 1-1 for pressure/temperature relationship). The vapor has now reached the 100% saturated vapor point and is at the lowest temperature it can be and still remain a vapor. This temperature is referred to as both its saturation and condensing temperature, and a pressure/temperature relationship exists. Any heat lost past the 100% saturated vapor point will gradually phase change the vapor to liquid (condensing). The heat removed from the vapor turning to liquid is referred to as latent heat and happens at a constant temperature of 100°F. As the vapor condenses to liquid, refrigerant molecules actually become more dense and get closer together. This molecular joining is what gives up most of the latent heat energy. This point is on the high side of the refrigeration system and is referred to as high pressure saturated vapor.

100% Saturated Liquid Point (Point #4)

Soon all of the vapor will give up its latent heat and turn to saturated liquid at a constant condensing temperature of 100°F. Any more heat given up by the refrigerant after this point will be sensible heat, because the phase change from vapor to liquid is complete. This point is still on the high side of the

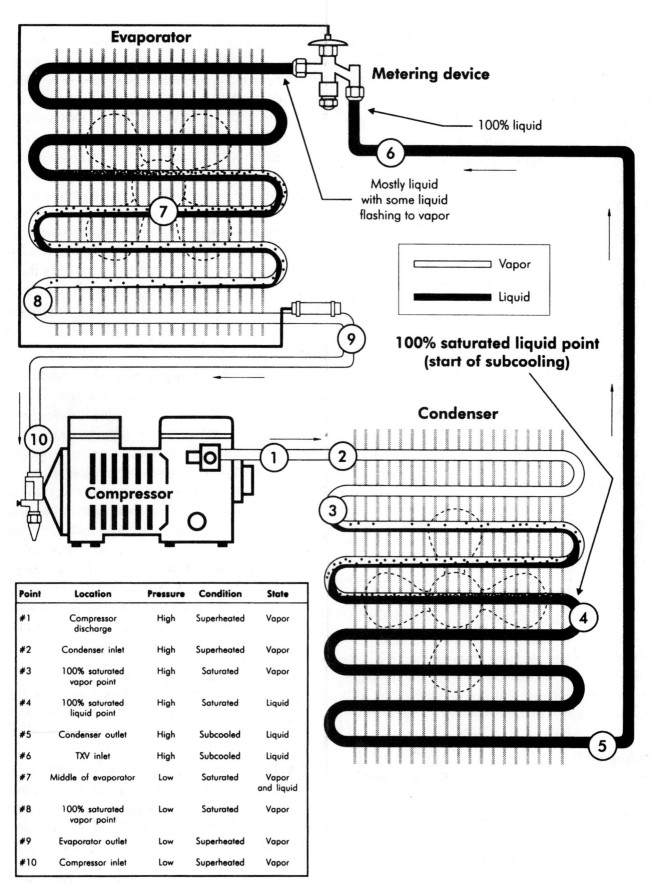

Point	Location	Pressure	Condition	State
#1	Compressor discharge	High	Superheated	Vapor
#2	Condenser inlet	High	Superheated	Vapor
#3	100% saturated vapor point	High	Saturated	Vapor
#4	100% saturated liquid point	High	Saturated	Liquid
#5	Condenser outlet	High	Subcooled	Liquid
#6	TXV inlet	High	Subcooled	Liquid
#7	Middle of evaporator	Low	Saturated	Vapor and liquid
#8	100% saturated vapor point	Low	Saturated	Vapor
#9	Evaporator outlet	Low	Superheated	Vapor
#10	Compressor inlet	Low	Superheated	Vapor

Figure 1-8. Basic refrigeration system showing refrigerant pressures, states, and condition locations

Accessories Found in Refrigeration and Air Conditioning Systems

---

refrigeration system and can be referred to as high pressure saturated liquid. The entire condensing process takes place between the 100% saturated vapor point and the 100% saturated liquid point. Any sensible heat lost past the 100% saturated liquid point is referred to as subcooling.

Condenser Outlet (Point #5)

Once the 100% saturated liquid point is reached in the condenser and more heat is removed, liquid subcooling occurs. Remember, any heat lost in the liquid past its 100% saturated liquid point is subcooling. Liquid subcooling can continue all the way to the entrance of the metering device if conditions are right. This point is on the high side of the system and is all subcooled liquid, so it is referred to as high pressure subcooled liquid. There is no pressure/temperature relationship at the subcooled condition, only at saturation. At a pressure of 124 psig, the corresponding temperature of the 100% saturated liquid point in the condenser is 100°F. The difference between 100°F and the condenser outlet temperature of 90°F is 10°F of condenser subcooling.

TXV Inlet (Point #6)

The inlet to the TXV is on the high side of the system and consists of subcooled liquid. This subcooling should continue from the 100% saturated liquid point in the condenser. The tubing from the condenser outlet to the TXV inlet is often referred to as the liquid line. The liquid line may be exposed to very high or low temperatures depending on the time of year. This will seriously affect whether or not subcooling takes place and to what magnitude. If the liquid line is exposed to hot temperatures, liquid line flashing may occur. Since this point is on the high side of the system and is subcooled liquid, it is referred to as high pressure subcooled liquid.

Middle of Evaporator (Point #7)

When the subcooled liquid enters the TXV, flashing of the liquid occurs. Once in the evaporator, the liquid refrigerant experiences a severe drop in pressure to the new saturation (evaporator) pressure of 6 psig. This pressure decrease causes some of the liquid to flash to vapor in order to reach the new

saturation temperature in the evaporator of 0°F. See Table 1-1 for the pressure/temperature relationship. Once this new evaporator temperature is reached, the liquid/vapor mixture starts absorbing heat from the product load and continues to change from liquid to vapor. This process happens at the new saturation temperature of 0°F, corresponding to the saturation (evaporator) pressure of 6 psig. This is a classic example of heat absorbed by the refrigerant without increasing in temperature, which is called the latent heat of vaporization. The heat energy absorbed in the refrigerant breaks the liquid molecules into vapor molecules instead of increasing its temperature. Since the refrigerant is both saturated liquid and vapor and is on the low side of the refrigeration system, it is referred to as a low pressure saturated liquid and vapor.

100% Saturated Vapor Point (Point #8)

Once all of the liquid changes to vapor in the evaporator, the 100% saturated vapor point is reached. This point is still at the evaporator saturation temperature of 0°F. Any more heat absorbed by the refrigerant vapor will result in a temperature rise of the refrigerant. This heat energy goes into increasing the velocity and spacing of the vapor molecules, because there is no more liquid to be vaporized. This increase in molecular velocity can be measured in degrees. Any heat added past this 100% saturated vapor point is superheat. Since this 100% saturated vapor point is in the low side of the system and is saturated vapor, it is referred to as low pressure saturated vapor.

Evaporator Outlet (Point #9)

The evaporator outlet temperature is used for evaporator superheat calculations. This point is located at the evaporator outlet next to the TXV remote bulb. Because it is located downstream of the 100% saturated vapor point, it is superheated. This point is in the low side of the refrigeration system and is referred to as low pressure superheated vapor. The difference between the 100% saturated vapor temperature of 0°F and the evaporator outlet temperature of 10°F is called evaporator superheat. In this example, there are 10°F (10°F – 0°F) of evaporator superheat.

Compressor Inlet (Point #10)

The compressor inlet consists of low pressure superheated vapor. This vapor feeds the compressor. As the refrigerant travels from the evaporator outlet down the suction line to the compressor, more superheat is gained. Superheat ensures that no liquid refrigerant enters the compressor at low evaporator loadings when TXV valves are known to lose control

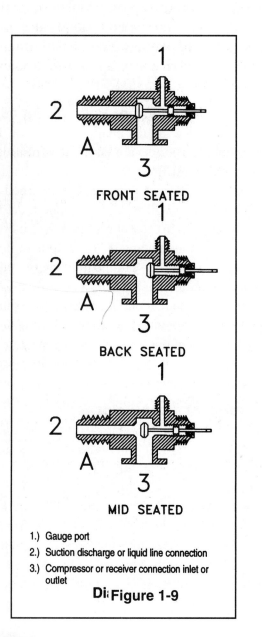

1.) Gauge port
2.) Suction discharge or liquid line connection
3.) Compressor or receiver connection inlet or outlet

Di Figure 1-9

of superheat settings. Because this point is superheated, no refrigerant pressure/temperature relationship exists.

The Gauge Manifold and Service Valves

The most important tool of the refrigeration serviceperson is the gauge and manifold set (Figure 1-22). It can be used for checking system pressures, recovering refrigerant, evacuating the system, adding oil, changing refrigerant and many other purposes.

Connecting the gauge manifold to a system is one of the most common service functions. To avoid introducing contaminates into the system, the hose connections must always be purged with refrigerant or evacuated before connecting the manifold. A consistent procedure should always be followed by the serviceperson in making connections. For an operating system containing refrigerant, proceed as follows:

First, back-seat the three way service valves to which gauges are to be connected so that the gauge ports are closed. Figure 1-9 shows three-way service valves in various service operations. If operating conditions are above 0 psig, tighten hose connections to both three-way service valves. Make sure common hose connection on manifold is open. Crack (open slightly) the high pressure manifold valve. Then crack the high pressure three-way service valve allowing refrigerant vapor pressure to push the air from the discharge and common hoses. Allow the air to bleed until eliminated. **This is considered a DeMinimus release of refrigerant.** Quickly close the high pressure valve on the manifold. Repeat the same procedure with the low pressure valves. The manifold is now connected to the system ready for use.

In case of systems where the low side pressure might be in a vacuum, all purging must be done from the high pressure three-way service valve. Back-seat the three-way service valves and tighten the hose connection to the system's high pressure service valve. Crack (open slightly) the high pressure

manifold valve. Leave the hose connection at low side service valve loose allowing refrigerant vapor pressure to push the air to the atmosphere. Make sure the common hose connection on the manifold is open. After a few seconds (do not allow refrigerant to escape) tighten the hose connection at the low pressure service valve, and then tighten the cap or plug in the common connection. Close the valves on the manifold, crack the low pressure three-way service valve and the manifold is then connected to the system ready for use.

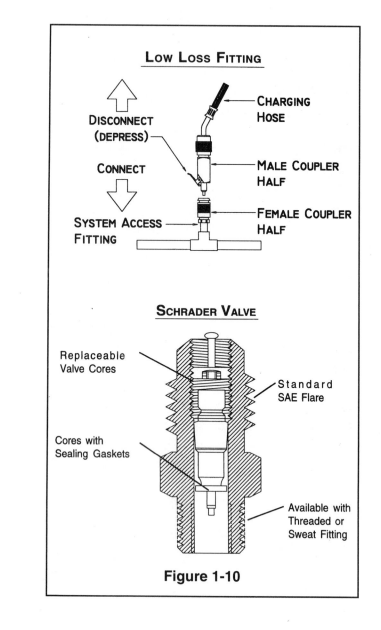

Figure 1-10

Removing Gauges with Three-way Service Valves

Bleeding high-side refrigerant left in hose into low side of system to minimize refrigerant losses has been used for many years. When the system has compressor service valves, the first step is to back-seat high-side service valve leaving low-side service valve mid-positioned. To prevent compressor overload, slowly open both high-side and low-side valves on manifold gauge set allowing high-side refrigerant caught in high-side hose into low side of system. Back-seat compressor's low-side compressor service valve. Close both high-side and low-side gauge manifold valves. There will be a minimum of refrigerant remaining in the manifold gauge and hoses. **This is considered a DeMinimus release of refrigerant emitted in the course of service procedures and is allowable by the EPA.**

Removing Gauges When Using the Quick Connect Fittings

When using the quick connect fittings available for use on Schrader valves, the method is very similar to units with service valves.

Remove high-side quick connect from Schrader valve. Being spring loaded, the quick connect will automatically seal. To prevent compressor overload, slowly crack open both high-side and low-side gauge manifold valves allowing refrigerant in high side to flow into the low-side of system. Once gauge pressures equalize, close both high-side and low-side gauge manifold valves. Disconnect the quick connect fitting from the Schrader valve on low side of system. It will automatically seal in the same manner as the high-side quick connect did. There will be a minimum of refrigerant remaining in the manifold gauge and hoses. Figure 1-10 shows both the Low Loss Fitting and Schrader Valve.

Suction Line Accumulator

Installed Between Evaporator and Compressor

Purpose: To Prevent Surges of Liquid Refrigerant from Reaching the Compressor

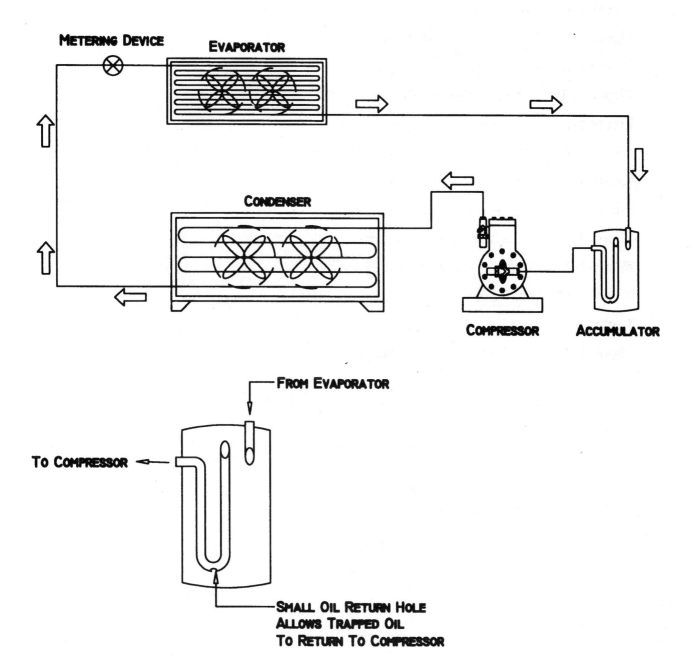

Figure 1-11. Suction Line Accumulator

Oil Separator

Installed in Hot Gas Line Between Compressor and Condenser

Purpose: To Prevent Excessive Oil from Circulating Throughout the System and to Return Oil Back to Compressor Crankcase

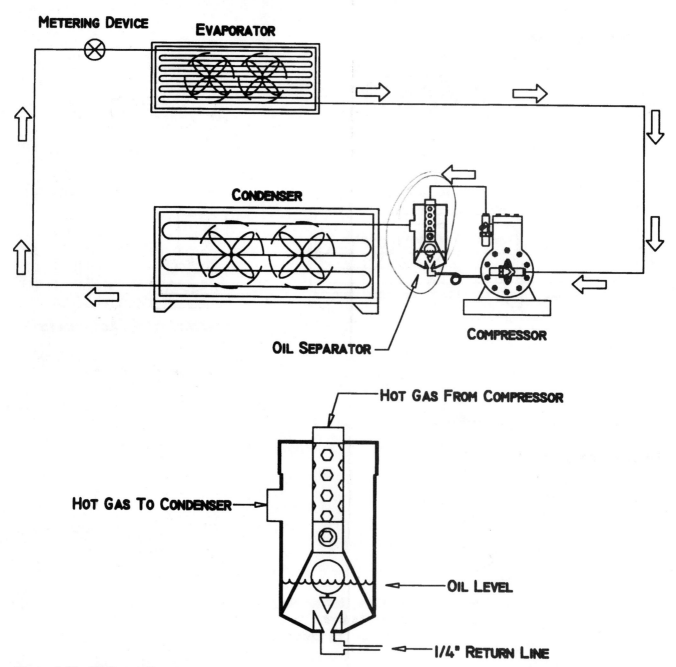

Figure 1-12. Oil Separator

Heat Exchanger

Installed Between Evaporator and Compressor when Using Capillary Tube System, the Capillary Tube will be Soldered Directly to the Suction Line

Purpose: To Subcool the Refrigerant in the Liquid Line and Superheat the Refrigerant in the Suction Line

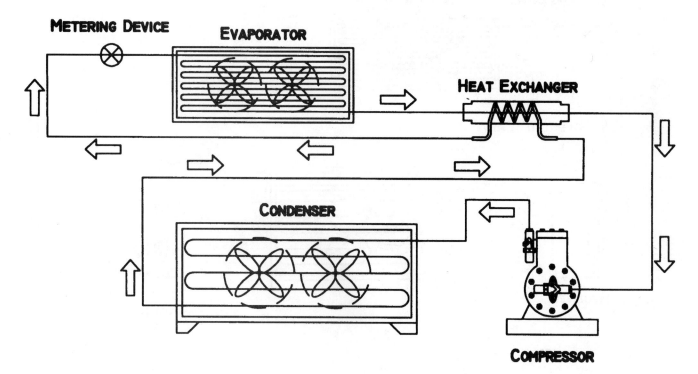

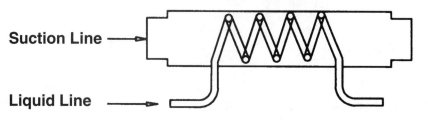

Figure 1-13. Heat Exchanger

Liquid Line Solenoid Valve

Installed Between Receiver and Evaporator

Purpose: To Allow for Automatic Pumpdown of a System

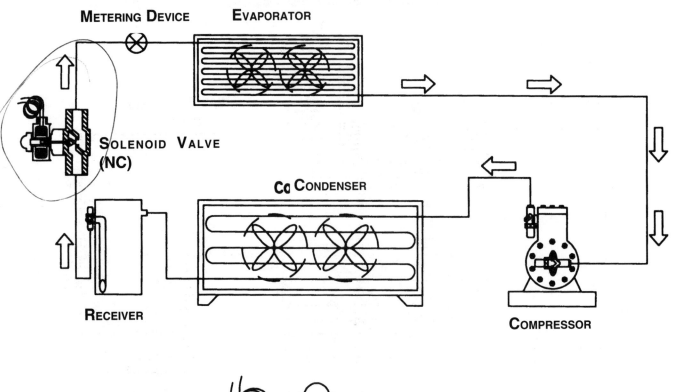

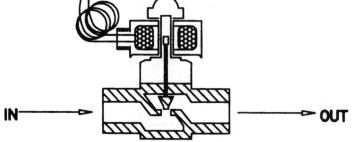

Normally Closed Solenoid Valve (NC)
Opens when Electrical Current is Applied

Figure 1-14. Liquid Line Solenoid Valve

Evaporator Pressure Regulator (EPR)

Installed in Suction Line Between Evaporator and Compressor

Purpose: Throttles Closed when Evaporator Load Drops Preventing Evaporator Pressure from Dropping Below a Set Point

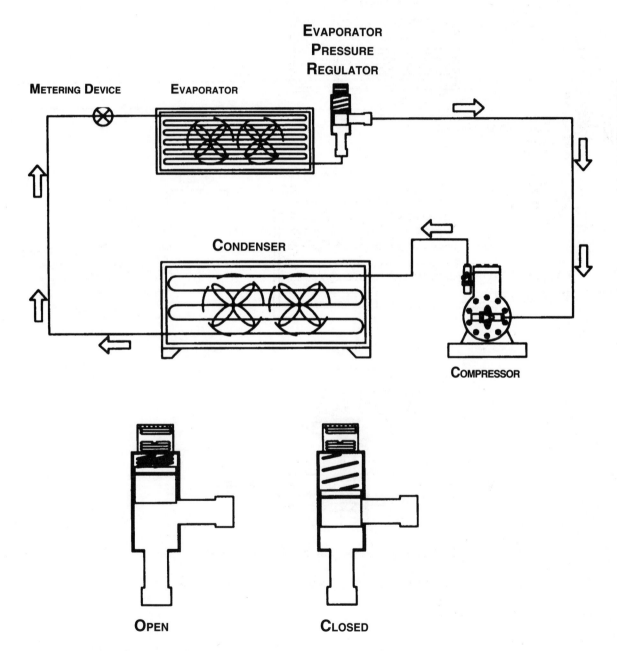

Figure 1-15. Evaporator Pressure Regulator (EPR)

Crankcase Pressure Regulator

(Holdback Valve)

Installed Between Evaporator and Compressor

Purpose: Used to Prevent Compressor Crankcase Pressure from Getting Too High. Throttles Closed When Crankcase Pressure Gets Too High

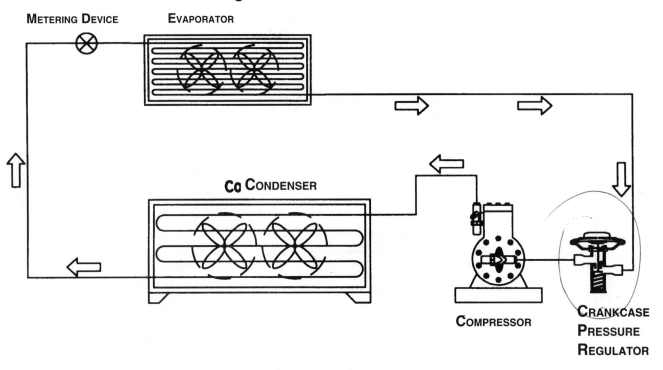

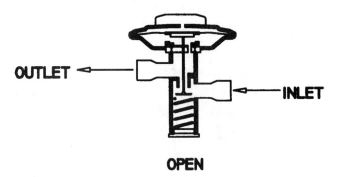

Figure 1-16. Crankcase Pressure Regulator (Holdback Valve)

Condenser Water Regulating Valve
(Used on Water Cooled Condensers)

Installed when Using a Water Cooled Condenser in Place of an Air Cooled Condenser

Purpose: To Regulate a Proper Flow of Cold Water to the Condenser to Reject Heat and Maintain a "Constant" Head Pressure

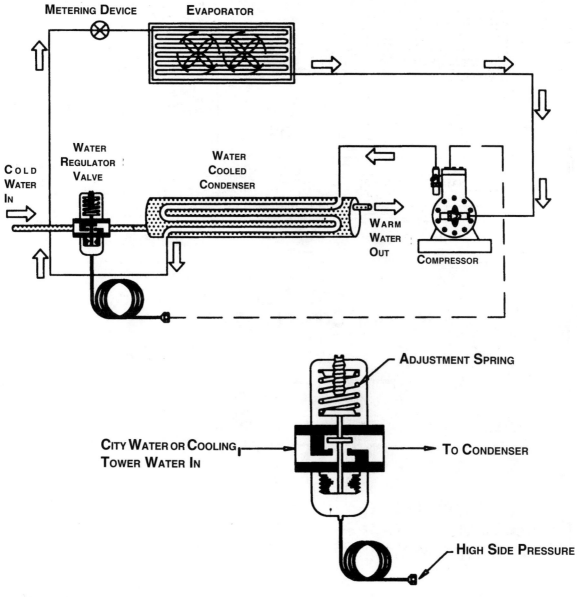

Figure 1-17. Condenser Water Regulating Valve

Four Way Reversing Valve
(Heating Mode Shown)
Installed in Heat Pumps

Purpose: To reverse the refrigerant cycle from cooling to heating

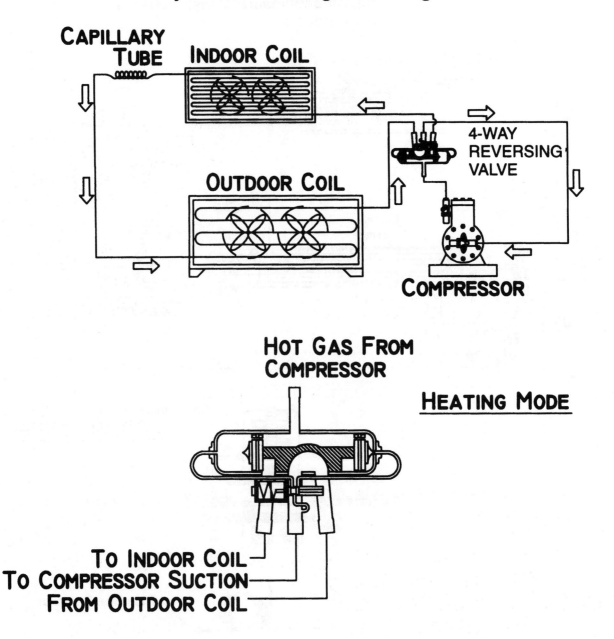

Figure 1-18. Four Way Reversing Valve (Heat Mode)

Four Way Reversing Valve (Cooling and Defrost Mode Shown) Installed in Heat Pumps

Purpose: To reverse the refrigerant cycle from heating to cooling

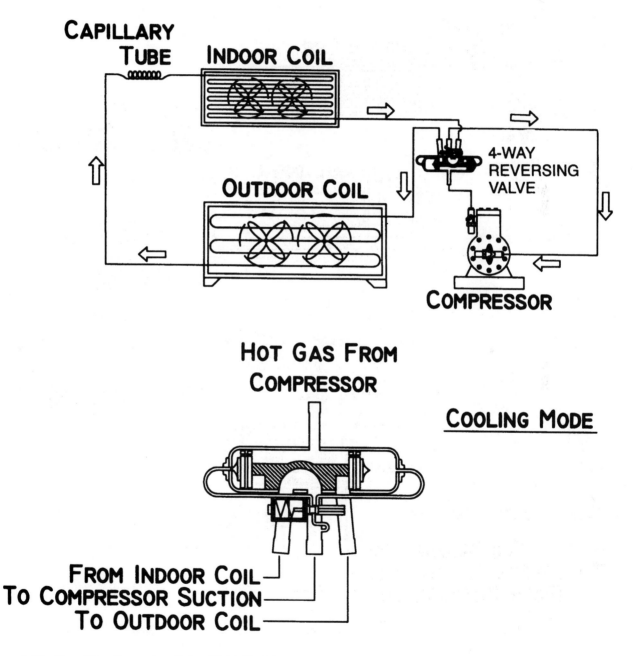

Figure 1-19. Four Way Reversing Valve (Cool Mode)

HEAT PUMP
FOUR WAY VALVE

Heat and Cool Using Two Thermostatic Expansion Valves and Two Bypass Lines with Check Valves

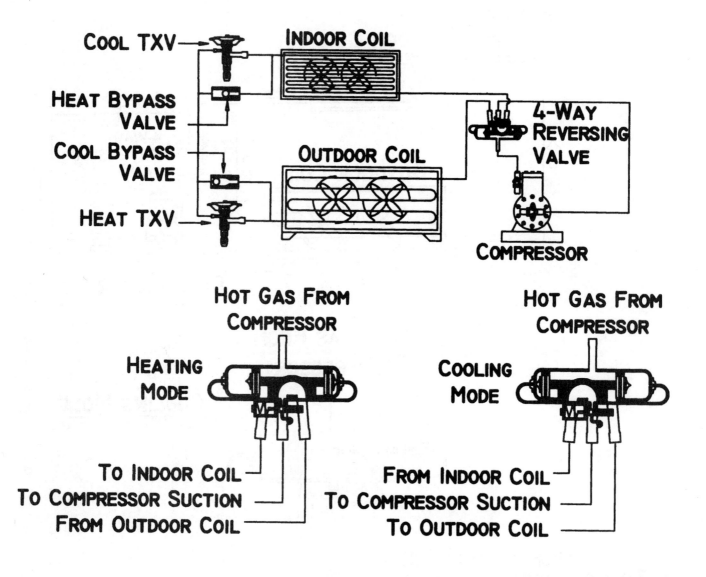

Figure 1-20. Four-Way Reversing Valve (Two TXVs)

GAUGE MANIFOLD

ADDING REFRIGERANT TO A SYSTEM WHILE IN OPERATION

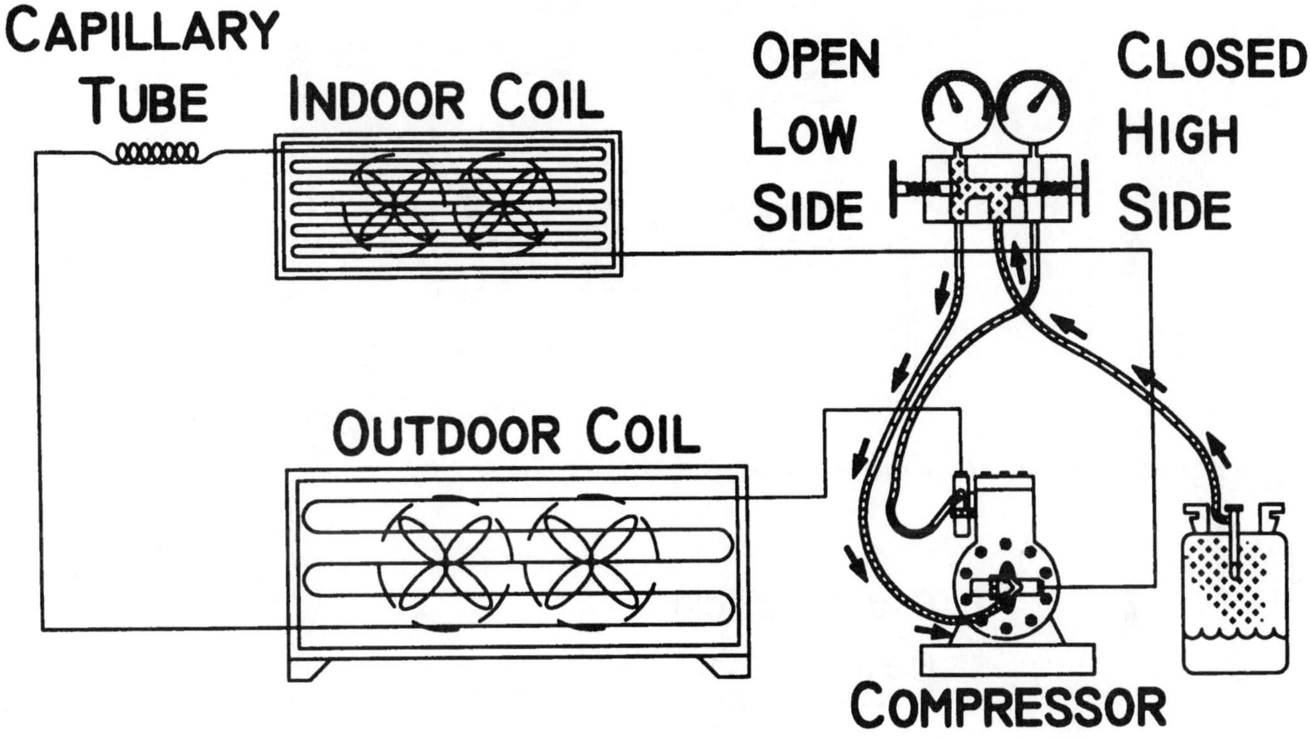

Note: Adding refrigerant vapor to the low side of a refrigeration system applies to refrigerants that do not have a temperature glide or fractionate (pure compounds). If the refrigerant has a temperature glide and fractionation potential, "liquid" must be "throttled" through the low side of the system. (Refer to Section Two on Blend Fractionation and Blend Temperature Glide.)

Figure 1-21. Gauge Manifold (Adding Refrigerant)

GAUGE MANIFOLD

Reading High and Low Side Pressure
While in Operation

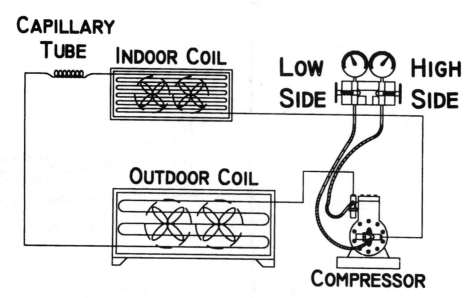

Figure 1-22. Gauge Manifold (High and Low Pressure)

Gauge Manifold

Evacuating Non-Condensibles
From a System (Unit Not Running)

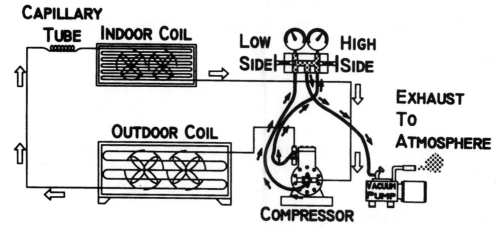

Figure 1-23. Gauge Manifold (Evacuation)

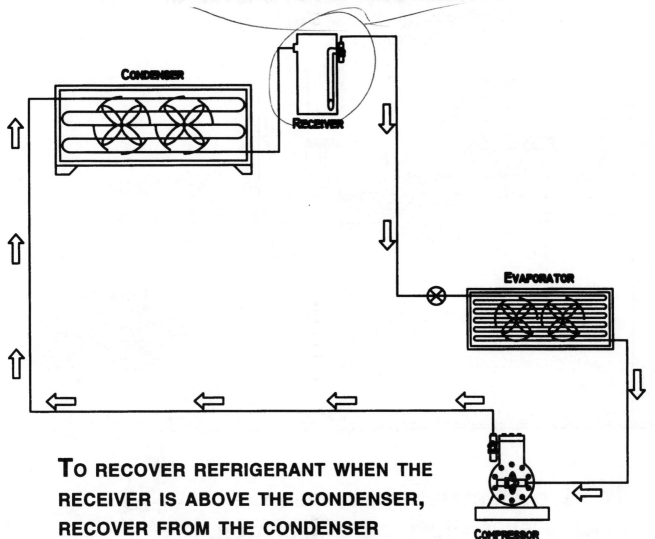

RECEIVER ABOVE CONDENSER

CONDENSER

RECEIVER

EVAPORATOR

COMPRESSOR

TO RECOVER REFRIGERANT WHEN THE RECEIVER IS ABOVE THE CONDENSER, RECOVER FROM THE CONDENSER OUTLET, RECEIVER OUTLET, AND METERING DEVICE TO THE EVAPORATOR INLET.

Figure 1-24. Receiver Above Condenser

CONDENSER ABOVE EVAPORATOR

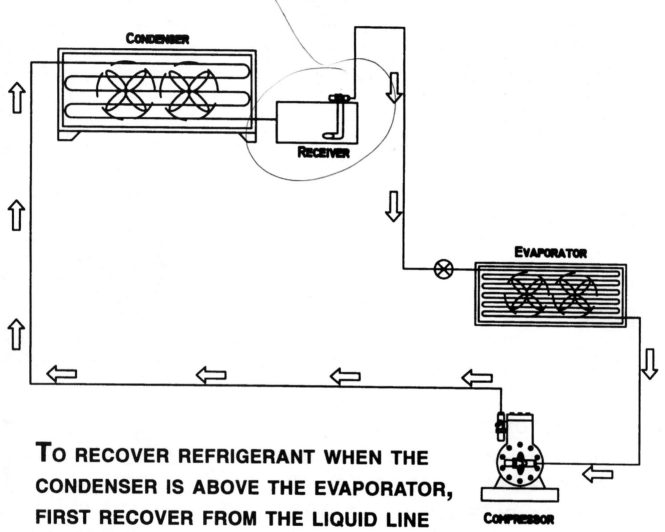

TO RECOVER REFRIGERANT WHEN THE CONDENSER IS ABOVE THE EVAPORATOR, FIRST RECOVER FROM THE LIQUID LINE ENTERING THE EVAPORATOR.

Figure 1-25. Condenser Above Evaporator

Notes

- ❑ Origin and Destination
- ❑ Molecular Structure
- ❑ CFCs, HCFCs, and HFCs
- ❑ Chlorofluorocarbons (CFCs)
- ❑ Hydrochlorofluorocarbons (HCFCs)
- ❑ Hydrofluorocarbons (HFCs)
- ❑ Refrigerant Blends
- ❑ Blend Fractionation
- ❑ Blend Temperature Glide
- ❑ Superheat and Subcooling Calculation Methods for Near-Azeotropic Blends
- ❑ Refrigerant Nomenclature

Origin and Destination

A refrigerant can be defined as any solid, liquid, or vapor that acts as a cooling agent by absorbing heat from another body or substance. Today, there is no one refrigerant that can be considered the "ideal" refrigerant. The many diverse cooling requirements and applications prevent an ideal refrigerant from existing. A refrigerant today not only must be economical to use, but physical, chemical, and thermodynamic properties must be satisfactory before it is put into use. Some of the more important properties that a refrigerant must have in today's refrigeration applications are listed below.

DESIRED REFRIGERANT PROPERTIES	
1.	Environmentally acceptable
2.	Non-toxic
3.	Non-flammable
4.	High latent heat of vaporization
5.	Chemically stable
6.	Material of construction compatible
7.	Lubricant soluble
8.	Low moisture solubility
9.	High dielectric strength
10.	Ease of transport handling
11.	Capable of being recycled
12.	Detectable at low concentrations
13.	Reasonable cost
14.	Readily available
15.	Field system charging capability

Molecular Structure

With the exception of ammonia, carbon dioxide, sulfur dioxide, and a few other refrigerants, most refrigerants originate mainly from two base molecules. These base molecules are methane and ethane. Both methane and ethane are pure hydrocarbons. Pure hydrocarbons are simply molecules that contain nothing but hydrogen and carbon in their structure (see diagram on page 34).

It wasn't until the 1920s that fluorocarbons were developed. Fluorocarbons, or fluorinated hydrocarbons, come from a family group called the halogenated hydrocarbons (or halocarbons). Halocarbons are hydrocarbons, like methane and ethane, which have had some or all of their hydrogen atoms replaced with fluorine, chlorine, bromine, astitine, or iodine atoms (see diagram on page 34). R-11, -12, -13, -14, -22, -32, -500, -502, -134a, 143a, -152a, -113, -114, -123, -124, and -125 are all halocarbons. Some or all of the hydrogen atoms of those refrigerants have been replaced with chlorine or fluorine. Chlorine, fluorine, and bromine are called halogens (see definition #1, page 35).

The term halocarbon comes from combining one of the halogens with a hydrocarbon like methane or ethane. Halogens that combine with the methane molecule are referred to as methane series halocarbons. Those that are combined with the ethane molecule are called ethane series halocarbons. These halocarbons, or halogenated hydrocarbons, are the structure for most of the refrigerants we use today.

CFCs, HCFCs, and HFCs

The way in which refrigerants are chemically structured has led to the use of acronyms when referring to the refrigerants. These acronyms are CFCs, HCFCs, and HFCs. All three of these refrigerant acronyms are explained in the following pages.

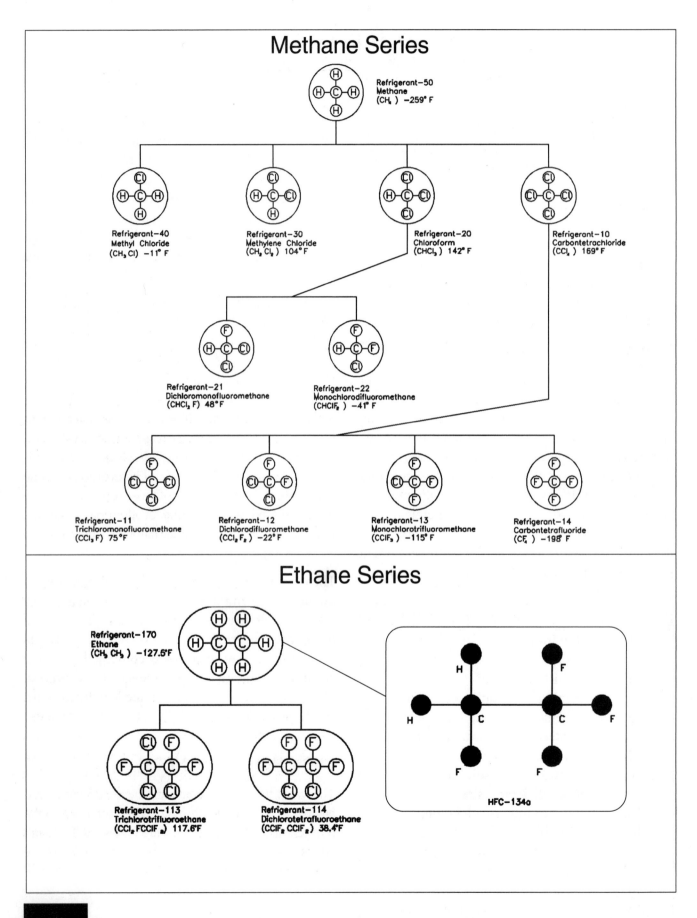

Methane Series

Refrigerant—50
Methane
(CH₄) —259° F

Refrigerant—40
Methyl Chloride
(CH₃Cl) —11° F

Refrigerant—30
Methylene Chloride
(CH₂Cl₂) 104° F

Refrigerant—20
Chloroform
(CHCl₃) 142° F

Refrigerant—10
Carbontetrachloride
(CCl₄) 169° F

Refrigerant—21
Dichloromonofluoromethane
(CHCl₂F) 48° F

Refrigerant—22
Monochlorodifluoromethane
(CHClF₂) —41° F

Refrigerant—11
Trichloromonofluoromethane
(CCl₃F) 75° F

Refrigerant—12
Dichlorodifluoromethane
(CCl₂F₂) —22° F

Refrigerant—13
Monochlorotrifluoromethane
(CClF₃) —115° F

Refrigerant—14
Carbontetrafluoride
(CF₄) —198° F

Ethane Series

Refrigerant—170
Ethane
(CH₃ CH₃) —127.5°F

Refrigerant—113
Trichlorotrifluoroethane
(CCl₂FCClF₂) 117.6°F

Refrigerant—114
Dichlorotetrafluoroethane
(CClF₂CClF₂) 38.4°F

HFC—134a

Chlorofluorocarbons (CFCs)

If all of the hydrogen atoms have been replaced in the base molecule by chlorine and fluorine, the refrigerant is said to be fully halogenated , and is referred to as a chlorofluorocarbon (CFC) (see Definition #1 below).   Refrigerants such as R-11, -12, -113, -114, -115, and -502 are phased out because of a manufacturing ban which started December 31, 1995. The diagram on page 34 shows that R-11, -12, and -13 of the methane series halocarbons are all chlorofluorocarbons (CFCs). They contain nothing but chlorine and fluorine atoms around the carbon atom. Also, both R-113 and R-114 of the ethane series halogens are CFCs. CFCs are very harmful to the atmosphere and have the highest ozone depletion potentials. CFCs have not been produced since 1995 because of their detrimental effect to the atmosphere. It is the chlorine in the CFC molecule that is very stable and will not break down for long periods of time. **CFCs are not water-soluble and do not wash out of the atmosphere with rain.** The chlorine molecule will reach the stratosphere and react with ozone, thus depleting the ozone layer. It is the ozone layer that protects the earth from the sun's harmful ultraviolet radiation.

DEFINITION #1: Halogens - *Any of the five non-metallic elements*
1. Fluorine   4. Iodine
2. Bromine   5. Astatine
3. Chlorine

Halogenate - *To cause some other element to combine with a halogen.*

Fully halogenated Hydrocarbon, or Chlorofluorocarbon (CFC) - *All hydrogen atoms in a hydrocarbon molecule are replaced with chlorine or fluorine.*
1. Atmospheric life of 75 to 120 years
2. High ozone depletion potential (ODP)
3. CFC-11, -12, -113, -114, -115 (and many more)

Hydrochlorofluorocarbons (HCFCs)

The base molecules of methane or ethane can be either fully or partially halogenated. With the CFCs, the molecule was fully halogenated because no hydrogen atoms remained. This gave CFCs a very stable configuration, thus a longer life. CFCs eventually will reach the stratosphere and deplete ozone.

Methane or ethane molecules may lose some of their hydrogen atoms to chlorine and fluorine, but still hang on to some hydrogen atoms. If this occurs, the molecule is said to be partially halogenated (see definition #2, below). Partially halogenated means that the halogens replace some of the hydrogens in either base molecule. Because some hydrogen atoms still exist, the entire molecule is referred to as a hydrochlorofluorcarbon (HCFC).   Some HCFCs used today are HCFC-22, -123, and -124.

DEFINITION #2: Partially Halogenated Hydrocarbon, or Hydrochlorofluorocarbon (HCFC) - *Some, but not all hydrogen atoms in a hydrocarbon molecule are replaced with chlorine or fluorine.*
1. Have shorter atmospheric lives
2. Less ozone depletion potential (ODP)
3. Referred to as HCFCs
4. HCFC-22, HCFC-124, HCFC-123

HCFCs are less detrimental to the atmosphere and most have much less ozone depletion and global warming potentials than CFCs. HCFC molecules contain chlorine, fluorine, and hydrogen atoms bonded to a carbon atom (see diagram on page 34). Because HCFCs do contain chlorine, they still damage the ozone layer if they reach it. However, because of the hydrogen bonds, these molecules are much less stable and have shorter lives. Therefore, not many HCFCs reach the stratosphere. Most HCFCs decompose in the lower atmosphere. Thus, they have a lower ozone depletion potential (ODP). They do have some global warming potential while in the lower atmosphere.

Hydrofluorocarbons (HFCs)

Often, a base molecule is halogenated by the fluorine atom only. There are no chlorine atoms anywhere in the molecule. The right side of the ethane series diagram on the bottom of page 34 shows the HFC-134a molecule. The base molecule of HFC-134a is ethane. Note that there are no chlorine atoms in the molecule. Only fluorine and hydrogen atoms surround the center carbon atoms. Because HFCs contain no chlorine, they have an ozone depletion potential of zero. However, they do have small global warming potentials. Several HFC refrigerants which have increased in popularity are HFC-32, -134a, -143a, -152a, and -125.

HFC-134a

HFC-134a is the alternate refrigerant of choice to replace CFC-12 and HCFC-22 in many medium and high temperature stationary refrigeration and air conditioning applications, and in automotive air conditioning. The reason for this is that the pressure/temperature relationship, and the latent heat values of HFC-134a are very similar to those of CFC-12. The automobile industry is using HFC-134a in place of CFC-12 because of its low hose permeability along with satisfactory efficiencies. HFC-134a suffers capacity losses when used as a low temperature refrigerant. Polarity differences between commonly used organic mineral oils and HFC refrigerants make HFC-134a insoluble, thus incompatible with mineral oils used in many refrigeration and air conditioning applications today. HFC-134a is not a drop-in refrigerant replacement.

HFC-134a systems must employ synthetic polyolester (POE) or polyalkylene glycol (PAG) lubricants (see chart next column). HFC-134a is also replacing CFC-12 and CFC-500 in many centrifugal and reciprocating chiller applications. HFC-134a is also being looked at to replace CFC-11 in new centrifugal chiller equipment. Ester-based lubricants are also required in these applications because of HFC-134a's incompatibility with organic mineral oils. In many centrifugal chiller applications,

efficiencies improved but not without some reductions in capacity.

HFC-134a / Oil Miscibility	
	HFC-134a
Polyalkylene Glycol Synthetic Oils	(G)
Polyolester Synthetic Oils	(G)
Naphthenic Mineral Oils	(P)
Paraffinic Mineral Oils	(P)
Alkylbenzene Synthetic Oils	(P)

(G)   Good miscibility
(P)   Poor miscibility

Aftermarkets

The service and retrofit markets also face the refrigerant and oil dilemma. Retrofitting an existing CFC-12/mineral oil system to a HFC-134a/polyolester system will require removing from 95 to 99 percent of the mineral oil. This could require at least three oil flushes with Polyol Ester oil, and is a labor intensive retrofit. Joint efforts between major compressor manufacturers and oil companies have published written procedures for this retrofit.

The automotive industry has used polyalkylene glycol (PAG) as the first generation lubricant with HFC-134a. PAGs are extremely hygroscopic and need a controlled handling environment. The automotive industry then moved to a Polyol Ester lubricant for use with HFC-134a because esters are less hygroscopic and will tolerate more chlorine contaminants for aftermarket retrofits and servicing. Also, Polyol Ester lubricants will not form peroxides and acids from incidental exposures to air.

Blends

In the meantime, widespread research is being done for a "drop-in" replacement for CFC-11, CFC-12, CFC-502, HCFC-22, and many other refrigerants. Near-azeotropic refrigerant mixtures (NARMs) are now being researched and manufactured by major chemical companies. NARMs, or blends of refrigerants, are being mathematically modeled with computers to tailor the refrigerant blend's characteristics to give maximum system efficiency and performance. Even vapor pressures can be adjusted by varying the percentage of each constituent in the blend. In fact, many refrigerant manufacturers will use the same blend, but vary the constituent percentages in the blend for use in different evaporating temperature applications. Percentage composition changes are also used to lower compression ratios and discharge temperatures for maximum operating performances and efficiencies.

Refrigerant blends can be HCFC based, HFC based, or a combination of both. Most refrigerant blends are either binary or ternary blends. **Binary blends consist of two refrigerants mixed together, where ternary blends consist of three refrigerants**. The HCFC based blends are only interim CFC replacements because of their chlorine content. Because HCFCs constitute a major percentage of some blends, the blends have lower ozone depletion and global warming potentials than most CFC and HCFC refrigerants that they are replacing. The HFC based blends will be long-term replacements for certain CFCs and HCFCs until researchers can find pure compounds to replace them.

HCFC-22 Replacement Candidates

For years, chemical companies have been researching refrigerants and refrigerant blends in order to find a permanent substitute for HCFC-22. HCFC-22 is scheduled for a total phase out in the year 2020 under the Montreal Protocol, with partial phase out starting sooner. Listed below are some of the HCFC-22 replacement blends.

R-410A is an azeotropic, binary refrigerant blend consisting of HFC-32 and HFC-125. This binary blend does not have a temperature glide and will not fractionate. R-410A is an HFC-based refrigerant blend, which is being used by original equipment manufacturers on new equipment in place of HCFC-22. It is a long-term replacement for HCFC-22 in commercial and residential air conditioning

applications. R-410A systems operate at much higher pressures (approximately 60% higher) than standard HCFC-22 systems. In fact, R-22 service equipment (hoses, manifold gauge sets, and recovery equipment) cannot be used on R-410A systems because of these higher operating pressures. Service equipment used for R-410A must be rated to handle higher operating pressures. Safety glasses and gloves should always be worn when working with R-410A. R-410A systems use synthetic Polyol Ester (POE) lubricant in their crankcases, and have higher efficiencies than standard R-22 systems. The tables on pages 38 and 39 list other HCFC-22 long-term and interim replacement refrigerants and blends.

R-407C is a near-azeotropic, ternary refrigerant blend consisting of HFC-32, HFC-125, and HFC-134a. R-407C is an HFC-based refrigerant blend with a high temperature glide and the ability to fractionate. It is a long-term replacement for HCFC-22 in commercial and residential air conditioning applications. R-407C is being used by original equipment manufacturers in new equipment and can also be used as a retrofit refrigerant blend. It has a very close capacity to R-22, but has lower efficiencies. R-407C systems use synthetic Polyol Ester lubricant in their crankcases.

CFC-502 Replacements

Because the Montreal Protocol dictates that CFC-502 (R-502) along with all other CFCs can no longer be manufactured as of the end of 1995, much research has been done to find a replacement refrigerant or refrigerant blend. Listed below are some CFC-502 replacement blends.

R-404A is a near-azeotropic, ternary refrigerant blend consisting of HFC-125, HFC-143a, and HFC-134a. R-404A has a small temperature glide and has the potential to fractionate. It is being used by original equipment manufacturers on new equipment and also on retrofitted equipment. R-404A systems use a polyol ester lubricant, and is a close match to CFC-502 in capacity and efficiency. R-404A is used as a medium and low temperature commercial refrigeration long-term replacement for CFC-502.

ALTERNATIVE REFRIGERANTS
Low-and Medium-Temperature Commercial Refrigeration
Long-Term Replacements

ASHRAE #	Trade Name	Manufacturer	Replaces	Type	Lubricant [a]	Applications	Comments
R-507 (125/143a)	AZ-50 507	AlliedSignal DuPont	R502 & HCFC-22	Azeotrope	Polyol Ester	New Equipment & Retrofits	Close match to R-502; higher efficiency than 404A; higher efficiency than R-22 at low temperature.
R-404A (125/143a/134a)	404A	AlliedSignal Elf Atochem	R-502 & HCFC-22	Blend (small glide)	Polyol Ester	New Equipment & Retrofits	Close match to R-502; higher efficiency than R-22 at low temperature.
	HP62	DuPont					
R-407D (32/125/134a)	407D	ICI	R-500	Blend (moderate glide)	Polyol Ester	New Equipment & Retrofits	Slightly higher capacity.
			R-12 Low Temp.				Higher capacity

Low-and Medium-Temperature Commercial Refrigeration
Interm Replacements [b]

ASHRAE #	Trade Name	Manufacturer	Replaces	Type	Lubricant [a]	Applications	Comments
R-402A (22/125/290)	HP80	AlliedSignal DuPont	R-502	Blend (small glide)	Alkylbenzene or Polyol Ester	Retrofits	Higher discharge pressure than R-502
R-402B (22/125/290)	HP81	AlliedSignal DuPont	R-502	Blend (small glide)	Alkylbenzene or Polyol Ester	Ice Machines	Higher discharge temperature than R-502
R-408A (125/143a/22)	408A	AlliedSignal DuPont Elf Atochem	R-502	Blend (small glide)	Alkylbenzene or Polyol Ester	Retrofits	Higher discharge temperature than R-502

Very Low-Temperature Commercial Refrigeration
Long-Term Replacements

ASHRAE #	Trade Name	Manufacturer	Replaces	Type	Lubricant [a]	Applications	Comments
R-23	HFC-23	AlliedSignal DuPont ICI	R-13	Pure Fluid	Polyol Ester	New Equipment & Retrofits	Higher discharge temperature than R-13
R-508B (23/116)	508B 95	AlliedSignal DuPont	R-13 & R-503	Azeotrope	Polyol Ester	New Equipment & Retrofits	
R-508A (23/116)	508A	ICI	R-13 & R-503	Azeotrope	Polyol Ester	New Equipment & Retrofits	

Medium-Temperature Commercial Refrigeration Long-Term Replacements

ASHRAE #	Trade Name	Manufacturer	Replaces	Type	Lubricant [a]	Applications	Comments
R-134a	HFC-134a	AlliedSignal DuPont Elf Atochem ICI	CFC-12	Pure Fluid	Polyol Ester	New Equipment & Retrofits	Close match to CFC-12

Courtesy: Honeywell

Medium-Temperature Commercial Refrigeration Interim Replacements[b]

ASHRAE #	Trade Name	Manufacturer	Replaces	Type	Lubricant [a]	Applications	Comments
R-401A (22/152a/124)	MP39	AlliedSignal DuPont	CFC-12	Blend (moderate glide)	Alkylbenzene or Polyol Ester or in some cases Mineral oil[d]	Retrofits [c]	Close to CFC-12. Use where evaporator temperature -10° F or higher
R-401B (22/152a/124)	MP66	AlliedSignal DuPont	CFC-12	Blend (moderate glide)	Alkybenzene or Polyol Ester	Transport[C] Refrigeration Retrofits	Close to CFC-12 Use where evap. temperature below -10° F
			R-500		Alkybenzene or Polyol Ester or in some cases Mineral Oil [d]	Retrofits including Air Conditioners & Dehumidifiers	
B-406A (22/142b/600a)	GHG	Peoples Welding Supply	CFC-12	Blend (high glide)	Mineral Oil or lkybenzene	Retrofits	Can segregate to flammable composition
R-409A (22/124/142b)	R-409A	AlliedSignal DuPont Elf Atochem	CFC-12	Blend (high glide)	Alkybenzene or Polyol Ester or in some cases Mineral Oil [d]	Retrofits[C]	Higher capacity then CFC-12 Similar to MP66
R-414A (pending) (22/124/142b/ 600a)	Autofrost	Peoples Welding Supply	CFC-12	Blend (high glide)	Alkybenzene or Polyol Ester or in some cases Mineral Oil [d]	Retrofits	Similar to 409A
R-414B (pending) (22/124/142b/ 600a)	Hot Shot	ICOR International	CFC-12	Blend (high glide)	Alkybenzene or Poly Ester or in some cases Mineral Oil [d]	Retrofits	Similar to 409A
R-416A (pending) (124/134a/600)	FRIGC FR-12	Intercool Energy	CFC-12	Blend (small glide)	Polyol Ester	Retrofits	Lower pressure then 134a at high ambient conditions.

Commercial and Residential Air-Conditioning Long-Term Replacements

ASHRAE #	Trade Name	Manufacturer	Replaces	Type	Lubricant [a]	Applications	Comments
R-123	HCFC-123	AlliedSignal DuPont	CFC-11	Pure Fluid	Alklbezene or Mineral Oil	Centrifugal Chillers	Lower capacity than R-11. With modifications, equivalent performance to CFC-11
R-134a	HFC-134a	AlliedSignal DuPont Elf Atochem ICI	CFC-12	Pure Fluid	Polyol Ester	New Equipment & Retrofits	Close match to CFC-12
			HCFC-22	Pure Fluid	Polyol Ester	New Equipment	Lower capacity than HCFC-22; larger equipment needed.
R-410A (32/125)	AZ-20 9100 410A	AlliedSignal DuPont Elf Atochem	HCFC-22	Azeotropic Mixture	Polyol Ester	New Equipment	Higher efficiency than HCFC-22. Requires equipment redesign.
R-407C (32/125/134a)	407C 9000	AlliedSignal Elf Atochem ICI Dupont	HCFC-22	Blend (high glide)	Polyol Ester	New Equipment &Retrofits	Lower efficiency than HCFC-22, close capacity to HCFC-22

(a) Check with the compressor manufacturer for their recommended lubricant.
(b) Interim replacements contain HCFCs, which are scheduled for phaseout under Montreal Protocol.
(c) Not recommended for automotive air-conditioning.
(d) For more information on when to use mineral oil, see Applications Bulletin GENAP1, Are Oil Changes Needed for HCFC Blends.

Disclaimer
All statements, information and data given herein are believed to be accurate and reliable, but are presented without guarantee, warranty or responsibility of any kind, expressed or implied. Statements or suggestions concerning possible use of products are made without representation or warranty that any such use is free of patent infrigement, and not recommendations to infringe any patent. The user should not assume that all safety measures are indicated, or that other measures are indicated, or that other measures may not be required.

R-402A and R-402B are also near-azeotropic, ternary refrigerant blends consisting of HCFC-22, HFC-125, and HC-290 with varying percentages. HC-290 is the hydrocarbon propane. These blends have small temperature glides and have the ability to fractionate. R-402A was designed as a retrofit refrigerant blend to replace CFC-502. It has a higher discharge temperature than R-502. R-402B is being used in some commercial ice machines and has a much higher discharge temperature than R-502. This makes for a more efficient hot gas defrost. Both R-402A and R-402B are medium and low temperature commercial refrigeration interim replacement blends. They are considered interim because of the HCFC-22 component in their composition. Refer to tables on page 38 and 39 for other CFC-502 long-term and short-term replacement refrigerants.

CFC-12 Replacements

As mentioned earlier, CFC-12 has a long-term high and medium temperature replacement refrigerant named HFC-134a. R-401A and R-401B are two near-azeotropic, ternary refrigerant blends consisting of HCFC-22, HFC-152a, and HCFC-124 with varying percentages for different applications. Both of these ternary blends are CFC-12 interim replacements in the medium temperature, commercial refrigeration market. Both blends have high temperature glides and have the potential to fractionate. R-401A and R-401B systems use either alkylbenzene or polyol ester synthetic lubricants in their crankcases. Both of their capacities and efficiencies are close to CFC-12 when evaporator temperatures are above -10°F. R-401A is a commercial refrigeration retrofit refrigerant, where R-401B is used for transportation refrigeration retrofits. Refer to the two tables on page 38 and 39 for other CFC-12 interim and long-term replacement refrigerants and refrigerant blends.

Some of these blends will be replacements for CFC-12 in certain temperature applications with minor retrofitting. Some major advantages of these blends are:

❏ Comparable capacity and efficiency when compared to CFC-12.

❏ Can use oils that are currently on the market (certain synthetic alkylbenzenes and esters).

❏ Are compatible with most materials of construction in today's systems.

CFC-11 Replacement

CFC-11, a low pressure refrigerant used in centrifugal chillers, has an interim retrofit replacement of HCFC-123. Alternative refrigerants for new chillers include ammonia, HCFC-123, HCFC-22, HFC-134a, and lithium bromide and water absorption systems. HCFC-123 has the advantage of lowest leak rates (emissions) due to its low pressure and high energy efficiency. The HCFC alternatives will be short-term candidates because of their chlorine contents.

All replacement refrigerant blends and pure compounds will require some sort of retrofit procedures often accompanied with oil changes. Please consult with the original equipment manufacturer (OEM) for specific retrofit procedures. Refer to the tables on pages 38 and 39 for alternative refrigerant nomenclature, replacement blends, lubrication types, and temperature applications for commercial refrigeration and commercial and residential air conditioning applications. The above mentioned tables cover both interim and long-term refrigerant replacements.

Blend Fractionation

Another important phenomenon of near-azeotropic and zeotropic refrigerant blends is **fractionation**. Fractionation is when one or more refrigerants of the same blend may leak at a faster rate than other refrigerants in the blend (see Definition #3, page 42). Fractionation is a change in composition of the blend by preferential evaporation of the more volatile components, or condensation of the less volatile components. Liquid and vapor must exist

Portions of definitions reprinted from ANSI/ASHRAE Standard 34-1989

simultaneously for fractionation to occur. This different leakage rate is caused from the different partial pressures of each constituent in the near-azeotropic mixture (see Definition #4, page 42). Fractionation also occurs because the blends are near-azeotropic mixtures and not pure compounds or pure substances like CFC-12 (see Definition #5, page 42). Fractionation was initially thought of as a serviceability barrier because the original refrigerant composition of the blend's constituents may change over time due to leaks and recharging. Depending on the blend's constituent make-up, fractionation may also segregate the blend to a flammable mixture if one or two constituents in the blend is flammable. When leaked, refrigerant blend fractionation may also result in faster capacity losses than single component pure compounds like CFC-12 or HCFC-22. However, further research proved that most blends were near-azeotropic enough for fractionation to be managed without flammability problems. The graph below shows a pure compound (CFC-12) and a near-azeotropic HCFC-based refrigerant blend consisting of HCFC-22/HFC-152a/HCFC-124. This HCFC-based blend (R-401A) is currently on the market and is an interim replacement for CFC-12 in medium and high temperature applications.

The graph shows pressure changes resulting from five vapor leaks from a confined container of CFC-12 (straight line), and the refrigerant blend (jagged lines). Each leak was a 50 percent weight loss

followed by a recharge of the original refrigerant or blend to their initial weights. Notice that CFC-12 did not lose any vapor pressure when leaked and recharged. This is because it is a pure compound with a single boiling point for any certain pressure. The blend, however, did lose some vapor pressure after each leak sequence. This is an example of blend fractionation. However, once recharged with the original blend to its initial weight, the vapor pressure did recover somewhat, but not quite to its original vapor pressure. After five consecutive leaks of 50 percent of its weight with recharging to its initial weight, the blend lost less than 10 percent of its original vapor pressure. This is considered satisfactory and will only slightly impact performance. One has to consider the severity of this leak/recharge test when trying to put an actual system in perspective.

Leaks of this magnitude should never occur because an environmentally conscious service technician should leak check HVACR systems before adding any refrigerant. It is illegal to top-off systems that significantly leak CFC, HCFC, or HFC-based refrigerants without first repairing the leak. Fines, costs of refrigerants, service callbacks, and damaged company reputation will deter technicians from intentionally topping-off leaky systems. As of November 15, 1995 it is illegal to intentionally vent HFC-based refrigerants, or to top-off systems that significantly leak HFC-based refrigerants.

To avoid fractionation, charging of a refrigeration system incorporating a near-azeotropic blend should be done with **liquid** refrigerant whenever possible. To ensure that the proper blend composition is charged into the system, it is important that liquid only be removed from the charging cylinder. Cylinders containing near-azeotropic blends are equipped with dip tubes, allowing liquid to be removed from the cylinder in the upright position. Once removed from the cylinder, these blends can be charged to the system as vapor, as long as all of the refrigerant removed is transferred to the system. When adding liquid refrigerant to an operating system, make sure liquid is <u>throttled</u>, thus vaporized,

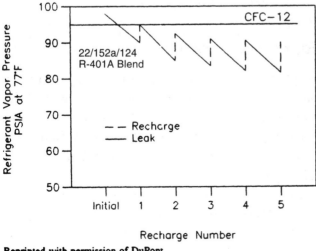

Reprinted with permission of DuPont

into the low side of the system to avoid compressor damage. A throttling valve can be used to ensure that liquid is converted to vapor prior to entering the system.

DEFINITION #3: Fractionation - *When one or more refrigerants of the same blend leak at a faster rate than other refrigerants in that same blend.*

DEFINITION #4: Azeotropic Mixture - *A mixture of two or more liquids, which, when mixed in precise proportion, act like a compound having a boiling temperature which is independent of the boiling temperatures of the individual liquids. The same numbers of molecules vaporize at the same rate. The liquid and vapor have the same composition. All of the mixture will vaporize before the temperature rises above the boiling point.*

Example: R-502 consists of R-22 and R-115
R-500 consists of R-12 and R-152a

Near-Azeotropic Mixture - *A blend which acts very similar to an azeotrope, however, it has a small volumetric composition change and temperature glide, or range of temperatures, as it evaporates and condenses.*

Example: R-401A, R-401B, R-402A, and R-402B

DEFINITION #5: Compound - *A substance formed by a union of two or more elements in definite proportions by weight. Only one molecule present. (For example: CFC-12, CFC-11, HCFC-22, HFC-134a, and water (H_2O).)*

Mixture - *A blend of two or more components which do not have a fixed proportion to one another and which, however thoroughly blended, are conceived of as retaining a separate existence. More than one molecule present. (For example: the near-azeotropic blends.)*

DEFINITION #6: Temperature Glide - *Range of condensing or evaporating temperatures for one pressure.*

Blend Temperature Glide

Near-azeotropic ternary blends have temperature glides when they evaporate and condense (see Definition #6). A pure compound like CFC-12, boils and condenses at a constant temperature for a given pressure. Since the blends are near-azeotropic, they will have some "temperature glide" or a range of temperatures in which they will boil and condense. The amount of glide will depend on system design and blend makeup. Temperature glide can range from 2 to 16° F. Since the saturated liquid temperature and the saturated vapor temperature for a given pressure are not the same, the constituent in the blend with the highest vapor pressure (lowest boiling point) will seek 100 percent saturated vapor before the other constituents. Sensible heat will now be gained by this refrigerant while the other constituents in the blend are still evaporating. This same phenomenon happens with condensation.

Some systems will not be affected by this temperature glide because it is design dependent. By all means, system design conditions must be evaluated when retrofitting with a blend. Because of the high percentage of HCFC-22 in some blends, the compressor may see higher condensing saturation temperatures and pressures when in operation. Because HCFC-22 has a relatively higher heat of compression when compared to other refrigerants, higher discharge temperature may be experienced. See tables on pages 38 and 39 for comments when using alternative refrigerants.

Superheat and Subcooling Calculation Methods for Near-Azeotropic Blends

As mentioned before, near-azeotropic blends are mixtures and not pure compounds, and have an associated temperature glide when they evaporate and condense. Temperature glide is nothing but a range of temperatures when evaporating or condensing for a given pressure. Pure compounds like CFC-12, HCFC-22, and HFC-134a, and azeotropic mixtures like CFC-502 and CFC-500 have only one associated temperature as they evaporate and condense for one given pressure. Because of temperature glide, the methods a service technician will calculate subcooling and superheat will be different for the near-azeotropic blends than with pure compounds and azeotropic blends.

Subcooling and Superheat Calculations with Temperature Glide

As an example, evaluate the condenser subcooling and evaporator superheat for a system incorporating R-401A (see diagram below). Assume an evaporating (low side) pressure of 5 psig (20 psia) and a condensing (high side) pressure of 105 psig (120 psia). The evaporator outlet temperature is 16°F and the condenser outlet temperature is 70°F.

Evaporator Superheat Calculation

Referring to the pressure/temperature relationship table for R-401A, one can see that there are two temperatures involved for the one pressure of 20 psia. The temperatures are -2° for the saturated vapor phase, and -13° for the saturated liquid phase. Pure compounds like CFC-12 only have one temperature for both liquid and vapor phases at a given pressure. Since evaporator superheat is calculated starting at the 100 percent saturated vapor point in the evaporator, we would use the temperature of -2° vapor for our superheat calculation, instead of the -13° liquid temperature which also corresponds to 20 psia.

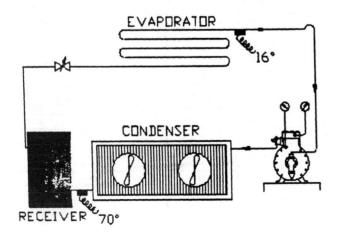

R-401A Blend Pressure/Temperature Relationships					
Saturated Vapor Pressure		Saturated Liquid Phase	Saturated Vapor Pressure		Saturated Liquid Phase
P, psia	T, F	P, psia	P, psia	T, F	P, psia
12.91	-20.00	17.10	70.76	60.00	83.83
13.25	-19.00	17.51	72.01	61.00	85.22
13.59	-18.00	17.93	73.27	62.00	86.63
13.95	-17.00	18.36	74.55	63.00	88.06
14.30	-16.00	18.80	75.85	64.00	89.50
14.67	-15.00	19.25	77.17	65.00	90.96
15.04	-14.00	19.70	78.51	66.00	92.44
15.42	-13.00	20.17	79.86	67.00	93.94
15.81	-12.00	20.64	81.23	68.00	95.46
16.21	-11.00	21.12	82.61	69.00	96.99
16.61	-10.00	21.61	84.02	70.00	98.55
17.02	-9.00	22.10	85.44	71.00	100.12
17.44	-8.00	22.61	86.88	72.00	101.71
17.87	-7.00	23.13	88.34	73.00	103.32
18.30	-6.00	23.65	89.82	74.00	104.94
18.74	-5.00	24.18	91.32	75.00	106.59
19.20	-4.00	24.73	92.83	76.00	108.26
19.66	-3.00	25.28	94.37	77.00	109.94
20.13	-2.00	25.84	95.92	78.00	111.65
20.60	-1.00	26.41	97.49	79.00	113.37
21.09	0.00	27.00	99.09	80.00	115.12
21.59	1.00	27.59	100.70	81.00	116.88
22.09	2.00	28.19	102.33	82.00	118.67
22.60	3.00	28.80	103.98	83.00	120.47
23.13	4.00	29.42	105.65	84.00	122.29
23.66	5.00	30.05	107.34	85.00	124.14
24.20	6.00	30.70	109.06	86.00	126.01
24.75	7.00	31.34	110.79	87.00	127.89
25.31	8.00	32.00	112.54	88.00	129.80
25.88	9.00	32.68	114.32	89.00	131.73
26.46	10.00	33.36	116.11	90.00	133.68
27.05	11.00	34.05	117.93	91.00	135.65
27.65	12.00	34.75	119.77	92.00	137.64
28.26	13.00	35.47	121.63	93.00	139.65
28.88	14.00	36.19	123.51	94.00	141.69
29.51	15.00	36.93	125.41	95.00	143.75
30.15	16.00	37.68	127.34	96.00	145.83
30.81	17.00	38.44	129.29	97.00	147.93
31.47	18.00	39.21	131.26	98.00	150.06
32.14	19.00	39.99	133.25	99.00	152.20

For example:

___ (16 degrees) ÷ evaporator outlet temperature

(-2 degrees) ÷ saturated vapor temperature at 20 psia (Refer to R-401A P/T table)

_____

(18 degrees) ÷ evaporator superheat

The evaporator is experiencing a temperature glide of 11°F. This is the difference between the liquid and vapor temperatures of -13° and -2° respectively. The evaporator is now vaporizing the refrigerant at a range of temperatures. The range is -13 to -2°F. This is much different than a pure compound like CFC-12 where at 20 psia both the liquid and vapor temperatures are -8°F (see CFC-12 Pressure/Temperature Relationship table). With CFC-12, it didn't matter if you chose the liquid or vapor temperature at 20 psia for a superheat calculation because they are the same temperature of -8°. However, with a refrigerant blend that has an associated temperature glide, choosing the wrong temperature for a given pressure will mean a superheat calculation error and possible ruined or inefficient equipment.

Condenser Subcooling Calculations

Referring to the Pressure/Temperature Relationship table for R-401A at 120 psia, once again you can see that there are two temperatures and two phases associated with a pressure of 120 psia. The phases are liquid and vapor and their temperatures are 83° and 92° respectively. Note that the vapor temperatures for a given pressure are always warmer than the liquid temperatures in both evaporator and condenser. Again, we have a temperature glide or a range of temperatures that the condenser is experiencing as its saturated vapor is condensing. Since condenser subcooling is calculated beginning at the 100 percent saturated liquid point in the condenser, the temperature of 83° saturated liquid temperature from the R-401A Pressure/Temperature table will be used for calculating condenser subcooling.

CFC-12 Pressure/Temperature Relationships

Temp. F	PSIA	PSIG	Temp. F	PSIA	PSIG
-40	9.3076	10.9709	70	84.888	70.192
-39	9.5530	10.4712	71	86.216	71.520
-38	9.8035	9.9611	72	87.559	72.863
-37	10.059	9.441	73	88.918	74.222
-36	10.320	8.909	74	90.292	75.596
-35	10.586	8.367	75	91.682	76.986
-34	10.858	7.814	76	93.087	78.391
-33	11.135	7.250	77	94.509	79.813
-32	11.417	6.675	78	95.946	81.250
-31	11.706	6.088	79	97.400	82.704
-30	11.999	5.490	80	98.870	84.174
-29	12.299	4.880	81	100.36	85.66
-28	12.604	4.259	82	101.86	87.16
-27	12.916	3.625	83	103.38	88.68
-26	13.233	2.979	84	104.92	90.22
-25	13.556	2.320	85	106.47	91.77
-24	13.886	1.649	86	108.04	93.34
-23	14.222	0.966	87	109.04	93.34
-22	14.564	0.270	88	111.23	96.53
-21	14.912	0.216	89	112.85	98.15
-20	15.267	0.571	90	114.49	99.79
-19	15.628	0.932	91	116.15	101.45
-18	15.996	1.300	92	117.82	103.12
-17	16.371	1.675	93	119.51	104.81
-16	16.753	2.057	94	121.22	106.52
-15	17.141	2.445	95	122.95	108.25
-14	17.536	2.840	96	124.70	110.00
-13	17.939	3.243	97	126.46	111.76
-12	18.348	3.652	98	128.24	113.54
-11	18.765	4.069	99	130.04	115.34
-10	19.189	4.493	100	131.86	117.16
-9	19.621	4.925	101	133.70	119.00
-8	20.059	5.363	102	135.56	120.86
-7	20.506	5.810	103	137.44	122.74
-6	20.960	6.264	104	139.33	124.63
-5	21.422	6.726	105	141.25	126.55
-4	21.891	7.195	106	143.18	128.48
-3	22.369	7.673	107	145.13	130.43
-2	22.854	8.158	108	147.11	132.41
-1	23.348	8.652	109	149.10	134.40
0	23.849	9.153	110	151.11	136.41
1	24.359	9.663	111	153.14	138.44
2	24.878	10.182	112	155.19	140.49
3	25.404	10.708	113	157.27	172.57
4	25.393	11.243	114	159.36	144.66
5	26.483	11.787	115	161.47	146.77
6	27.036	12.340	116	163.61	148.91
7	27.597	12.901	117	165.76	151.06
8	28.167	13.471	118	167.94	153.24
9	28.747	14.051	119	170.13	155.43
10	29.335	14.639	120	172.35	157.65
11	29.932	15.236	121	174.59	159.89
12	30.539	15.843	122	176.85	162.15
13	31.155	16.459	123	179.13	164.43
14	31.780	17.084	124	181.43	166.73
15	32.415	17.719	125	183.76	169.06

For example:

$$\frac{(83 \text{ degrees})}{(70 \text{ degrees})} \quad \div \quad \text{saturated liquid temperature at } 120 \text{ psia}$$
(Refer to R-401A P/T table)

(70 degrees) ÷ condenser outlet temperature

(13 degrees) ÷ condenser subcooling

The condenser is now experiencing a temperature glide of 9°, which is the difference between the saturated liquid temperature of 83° at 120 psia, and the saturated vapor temperature of 92° at 120 psia. The condenser would then have a condensing temperature range from 83 to 92°F. This is much different from CFC-12 which at 120 psia would have the same vapor and liquid temperature of approximately 93.5°F (see CFC-12 Pressure/Temperature relationship table on page 44). With CFC-12, it wouldn't matter if you chose the liquid or vapor temperature for a subcooling calculation because both of them are the same temperature. One must choose the saturated liquid temperature when calculating subcooling amounts for systems using a blend with an associated temperature glide. Manufacturers have now devised pressure/temperature charts to where it is near impossible to choose the wrong temperature for a given pressure. This is because when technicians are figuring superheat values, the chart instructs them to use the DEW POINT values only. When technicians are determining subcooling amounts, the chart instructs them to use BUBBLE POINT values only. The Pressure/Temperature Chart to the right illustrates this concept.

Blend Lubricants

The main lubricant for the HCFC-based blends will be a synthetic oil called alkylbenzene. One of the more popular alkylbenzenes has been marketed under the tradename of "Zerol." The blends are soluble in a mixture of alkylbenzene and mineral oils in concentrations of up to 20 percent mineral oil. This will make retrofitting a CFC/mineral oil system to a blend/alkylbenzene system possible without

PRESSURE-TEMPERATURE CHART

Note: In the original chart the vertical labels "BUBBLE POINT" and "DEW POINT" are printed within each refrigerant column, marking (with a pointing-hand symbol) the pressure range in which the listed value is to be read as the bubble point (for subcooling) or the dew point (for superheat). In the glide range some refrigerants show two numbers (bubble/dew), shown below as "x / y".

PSIG	Pink MP39 / 401A	Sand HP80 / 402A	Orange HP62 / 404A	Green KLEA 60 / 407A	Lt. Brown 9000 or KLEA 66 / 407C	Brown FX-56 / 409A
5*	-23	-59	-57	-45	-40	-22
4*	-22	-58	-56	-43	-39	-20
3*	-20	-56	-54	-42	-37	-19
2*	-19	-55	-53	-41	-36	-17
1*	-17	-54	-52	-39	-35	-16
0	-16	-53	-51	-38	-34	-15
1	-13	-53	-48	-36	-31	-12
2	-11	-48	-46	-33	-29	-9
3	-9	-45	-43	-31	-27	-7
4	-6	-43	-41	-29	-24	-5
5	-4	-41	-39	-27	-22	-2
6	-2	-39	-37	-25	-20	0
7	0	-37	-35	-23	-18	2
8	2	-36	-33	-21	-17	4
9	4	-34	-32	-20	-15	6
10	6	-32	-30	-18	-13	8
11	8	-30	-28	-16	-12	9
12	9	-29	-27	-15	-10	11
13	11	-27	-25	-13	-8	13
14	13	-26	-23	-12	-7	14
15	14	-24	-22	-10	-5	16
16	16	-23	-20	-9	-4	17
17	17	-21	-19	-8	-3	19
18	19	-20	-18	-6	-1	20
19	20	-19	-16	-5	0	22
20	21	-17	-15	-4	1	23
21	23	-16	-14	-2	3	25
22	24	-15	-12	-1	4	26
23	25	-14	-11	0	5	27
24	27	-12	-10	1	6	29
25	28	-11	-9	2	8	30
26	29	-10	-8	4	9	31
27	30	-9	-6	5	10	32
28	32	-8	-5	6	11	34
29	33	-7	-4	7	12	35
30	34	-6	-3	8	13	36
31	35	-5	-2	9	14	37
32	36	-4	-1	10	15	38
33	37	-2	0	11	16	39
34	38	-1	1	12	17	40
35	39	0	2	13	18	41
36	40 / 30	0	3	14	19	43
37	42 / 31	1	4	15	20	44
38	43 / 32	2	5	16	21	45 / 30
39	44 / 33	3	6	17	22	46 / 31
40	45 / 34	4	7	18	23	47 / 32
42	46 / 36	6	8	19	25	48 / 34
44	48 / 38	8	10	21	28	50 / 36
46	50 / 40	10	12	23	28	38
48	42	11	14	24	30	39
50	44	13	16	26	31	41
52	45	14	17	28	33	43
54	47	16	19	29	36	45
56	49	18	20	31	36	46
58	50	19	22	32	37	48
60	52	20	23	33	39	50
62	53	22	25	35	40	51
64	55	23	26	36	42 / 30	53
66	56	25	27	38	43 / 32	54
68	58	26	29	39	44 / 33	56
70	59	27	30 / 29	40 / 30	46 / 34	57
72	61	29	32 / 31	41 / 31	47 / 36	58
74	62	30	33 / 32	43 / 32	48 / 37	60
76	64	31	34 / 34	44 / 34	49 / 38	61
78	65	32 / 30	35 / 34	45 / 35	39	63
80	66	34 / 31	37 / 36	46 / 36	41	64
85	69	37 / 34	40 / 39	49 / 42	44	67
90	73	40 / 37	42 / 42	42	46	70
95	76	42 / 40	45 / 44	45	49	73
100	78	45 / 43	48 / 47	47	52	76
105	81	48 / 45	50	50	54	79
110	84	50 / 48	52	53	57	82
115	87	50	55	55	59	84
120	89	53	57	57	62	87
125	92	55	59	60	64	89
130	94	57	62	62	66	92
135	96	60	64	64	69	94
140	99	62	66	66	71	96
145	101	64	68	68	73	99
150	103	66	70	70	75	101
155	105	68	72	72	77	103
160	108	70	74	74	79	105
165	110	72	76	76	81	107
170	112	74	78	78	82	109
175	114	75	80	80	84	111
180	116	77	82	81	86	113
185	117	79	83	83	88	115
190	119	81	85	85	90	117
195	121	82	87	87	91	119
200	123	84	88	88	93	121
205	125	86	90	90	95	123
210	127	87	92	91	96	124
220	130	91	95	94	99	128
230	133	94	98	97	102	131
240	136	97	101	100	105	134
250	140	99	104	103	108	137
260	143	102	107	106	111	141
275	147	106	111	110	115	145
290	151	110	115	114	119	149
305	155	114	118	117	123	153
320	159	118	122	121	126	157
335	163	121	126	124	130	161
350	167	125	129	128	133	165
365	170	128	132	131	137	169

*Inches mercury below one atmosphere

Courtesy: Sporlan Valve Co.

45

extensive oil flushing. The only system change may be a quick oil flush and a different filter drier. In many of the retrofitted applications, the same thermostatic expansion device will be permitted. Some compressor manufacturers are using a mixture of mineral oil and alkylbenzene lubricants in their compressors for blend applications. Different applications and designs will dictate what lubricants to incorporate in each system. Most HFC-based blends will incorporate Polyol Ester lubricants. Retrofit guidelines have been written, and the original equipment manufacturer should be contacted before retrofitting to a blend.

Refrigerant Nomenclature

DuPont Chemicals devised a way to refer to refrigerants by numbers instead of their complex chemical names. The method is explained below.

First number on the right = the number of fluorine atoms.

Next number to the left = the number of hydrogen atoms plus one.

Next number to the left = number of carbon atoms minus one. (If equal to zero, omit).

The number of chlorine atoms in the compound is found by subtracting the sum of fluorine, bromine, and hydrogen atoms from the total number of atoms that can be connected to the carbon atoms.

EXAMPLE #1:

CFC-12
(dichlorodifluoromethane) CCl_2F_2

# of fluorine atoms	= 2
# of hydrogen atoms + 1	= 1
# of carbon atoms - 1	= 0 (omit)

Thus CFC-12

EXAMPLE #2:

HCFC-124
(monochlorotetrafluoroethane)
$CHClFCF_3$

# of fluorine atoms	= 4
# of hydrogen atoms + 1	= 2
# of carbon atoms - 1	= 1

Thus HCFC-124

EXAMPLE #3:

HFC-134a   (tetrafluoroethane) CF_3CH_2F

# of fluorine atoms	= 4
# of hydrogen atoms + 1	= 3
# of carbon atoms - 1	= 1

Thus HFC-134a

With knowledge of CFCs, HCFCs, and HFCs, and also the numbering system that follow these acronyms, let's dig a bit deeper and find out what the letters at the end of the numbering system mean using HFC-134a as an example.

We know that the molecule HFC-134a is a hydrofluorocarbon (HFC) and contains hydrogen, fluorine, and carbon. We also learned that the molecule contains 4 fluorine atoms, 2 hydrogen atoms, and 2 carbon atoms according to DuPont's numbering system.

The letter "a" at the end of the numbering system represents how symmetrical the molecular arrangement is. Many times two or more chemical compounds contain the same number of atoms of the same elements, but differ in structural arrangement or symmetry relative to perfect symmetry. When this is the case, the molecules are said to be isomers (see definition #7 on page 47). The HFC-134 molecule has two isomers. The two isomers are HFC-134 and HFC-134a.

Notice in the diagram on the bottom of page 47 that the HFC-134 molecule is very symmetrical and evenly distributes its fluorine and hydrogen atoms around the carbon atoms. In fact, each carbon atom of the HFC-134 molecule has exactly two fluorine and one hydrogen atoms. This arrangement is the most symmetrical arrangement.

DEFINITION #7: Isomer - *Any of two or more chemical compounds containing the same number of atoms of the same elements, but arranged differently.*

Example: HFC-134 and HFC-134a

However, the HFC-134a molecule is an isomer. It still has the same number of fluorine and hydrogen atoms around the carbon atoms, but they are arranged very differently. In fact, they are not arranged as symmetrical as the HFC-134 molecule. A closer look shows that there are three fluorine atoms around one carbon atom, and only one fluorine atom with two hydrogen atoms around the other carbon atom. This is an unsymmetrical arrangement relative to the HFC-134 molecule, so a letter "a" is placed after the numbering system to make it HFC-134a. This means that this isomer is next in line to HFC-134 in symmetry. If there was another isomer of this compound, it could be called HFC-134b meaning that its symmetry varies even farther from the perfect symmetry of HFC-134. Even though isomers of the same compound have the same number of atoms of the same elements, their structural arrangements and physical properties vary considerably.

Refrigerant Blend Nomenclature

Refrigerant blends are designated by their refrigerant numbers and weight proportions. The refrigerants will be listed first in order of increasing boiling points, followed by their respective weight percentages.

For example:

A blend of 40 percent wt. R-134a and 60 percent wt. R-12 will be designated by R-12/134a (60/40) or Refrigerant 12/134a (60/40). R-12 was listed first because it has the lower boiling point of the two refrigerants.

The blends also have refrigerant "R" numbers:

- The 400 series blends represent the near-azeotropic refrigerant blends (temperature glide and fractionation).

- The 500 series blends represent the azeotropic blends (negligible temperature glide and fractionation).

For example:

R-401 would indicate that the blend is a near-azeotrope, and the 1 would indicate that it is the first 400 series blend commercially produced.

R-502 would indicate that the blend is an azeotrope, and the 2 would indicate that it is the second 500 series blend commercially produced.

Often, refrigerant blends have the same R-number designation, but will be followed by different capital letters. Examples are, R-401A and R-401B, R-402A and R-402B, R-410A and R-410B, and lastly R-407A and R-407C. The different capital letters are used to distinguish between different percentages of the same constituents in each blend. For example, R-401A and R-401B are both ternary blends made up of the same three refrigerants (HCFC-22, HFC-152a, and HCFC-124). However, R-401A and R-401B each have different percentages of these same three refrigerants, thus the different capital letters at their ends.

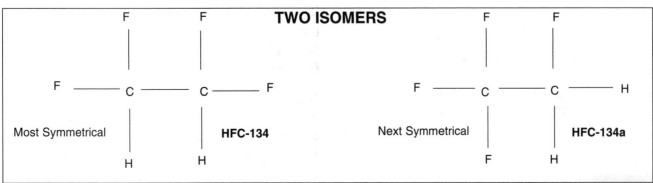

Notes

SECTION THREE
Refrigeration Oils and Their Applications

- ❑ Oil Properties
- ❑ Oil Groups
- ❑ Mineral Oils
- ❑ Synthetic Oils
- ❑ Alkylbenzene
- ❑ Glycols
- ❑ Esters
- ❑ Oil Additives
- ❑ Waste Oil

Today, oils used in refrigeration compressors are not considered standard lubricants by any means. Years of testing and research have made a science out of refrigeration oils, thus categorizing them as specialty products. Understanding the behavior of refrigeration oil requires information on composition, properties, and application.

In any refrigeration system, oil and refrigerant are always present. The refrigerant is the working fluid and is required for cooling. The oil's main purpose is to lubricate the compressor. Refrigerant and oil are miscible (mixable) in one another and their magnitude of miscibility will depend on the type of refrigerant, the temperature, and the pressure both are exposed to. A certain amount of oil will always leave the compressor's crankcase and be circulated with the refrigerant. Refrigerant and oil can separate into two phases and become immiscible in one another at certain temperatures. Refrigeration oils and refrigerants are often miscible in one another over wide temperature ranges. If not soluble, the oil would not move freely around the system and oil rich pockets would form causing restrictions, poor heat transfer, and inadequate oil return to the compressor.

Even though the primary function of an oil is to minimize mechanical wear through lubrication and to reduce the effects of friction, oil in a refrigeration system accomplishes many more tasks. Oil acts as the seal between the discharge and suction sides of the compressor. Oil will prevent excessive blow-by around the piston in a reciprocating compressor. Oil

also prevents blow-by in some centrifugal compressors by putting a seal around its vanes. Oil acts as a noise dampener reducing internal mechanical noise within a compressor. Oil also performs heat transfer tasks by sweeping away any heat from internal rotating and stationary parts.

Bearing performance is affected by both boundary and hydrodynamic lubrication. Boundary lubrication can be seen in the compressor's reciprocal motion

Refrigeration Oil Functions
• Minimizes mechanical wear
• Reduces friction
• Lubricates
• Seals - prevents blow-by
• Deadens noise
• Transfers heat - "cools"

especially when a compressor starts. Metal rubbing against metal during start-ups is an example of boundary lubrication. Load carrying ability or film strength are terms often associated with boundary lubrication. Boundary lubrication can also be seen during high load situations and is undesirable. Hydrodynamic lubrication, on the other hand, involves separating bearing elements with an oil film while they are moving. Theoretically, mechanical parts in an operating compressor should never touch one another. In fact, lubrication is defined as the separation of moving parts by an oil film. There should always be an oil film dividing or supporting compressor parts to prevent wear. Hydrodynamic lubrication depends on the viscosity or thickness of the oil, and involves the oil passing load tests.

Oil Properties

The fact that refrigeration oil comes in direct contact with the working refrigerant and mixes with it makes it very important that refrigeration oil be specially designed and formulated to handle this function. Listed below are definitions and discussions of some of the most important properties that must be considered when designers select an oil for a specific application.

1. **Viscosity** - The resistance that the oil offers to flow. High viscosity means a thick oil, and low viscosity a thin oil. Also, defined as the body of the oil or its ability to perform a lubricating function. Sometimes referred to as the thickness of the oil. Viscosity of an oil is measured in Saybolt Universal Seconds (SUS). You will see either (SUS) or (SSU) in manufacturer's literature depending on preference. SUS or (SSU) is nothing but a "time" in seconds that it takes for a known sample of oil, at a controlled temperature (usually 100°F), to flow by gravity from a container through a capillary tube of known length and diameter. If an oil at 100° F took 150 seconds to flow through the capillary tube, the oil would have a viscosity of 150 SUS. If it took 300 seconds, the oil would have a viscosity of 300 SUS. The SUS value is usually always printed on the oil's container. The more time it takes for the oil to flow through the tube, the thicker it must be, and thus has a higher viscosity. The higher the SUS value, the higher the viscosity and the thicker the oil. Viscosity is affected by temperature and by the amount of refrigerant dissolved in the oil (see chart below).

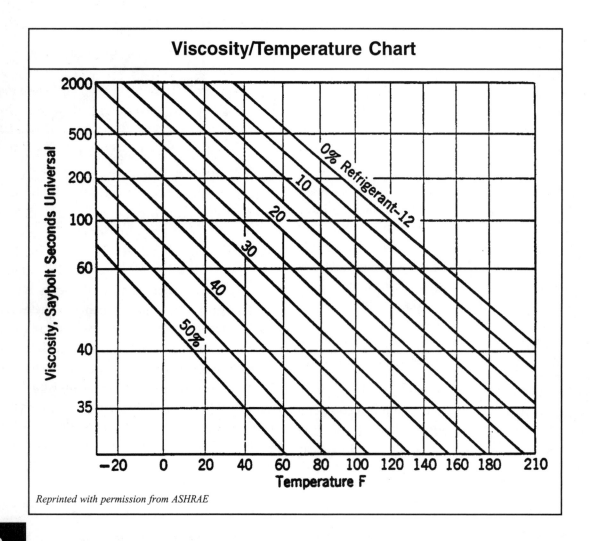

Reprinted with permission from ASHRAE

Refrigerant Transition and Recovery Certification Program Manual

The top line of the viscosity temperature chart represents pure oil and how its viscosity decreases as temperature increases. The other lower lines represent percentages of R-12 mixed with the oil. For a constant temperature, as the percentage of refrigerant increases, the viscosity decreases or a thinning of the oil results. Thus, both increases in temperature and percentage of refrigerant in the oil decrease the viscosity of most oils.

2. **Chemical stability** - The oil's ability to lubricate for extended periods of time without breaking down or reacting with other materials of construction. The original oil may often remain in a compressor for over 10 years. Also, reactions among oil, refrigerant, and oxygen can cause problems such as copper plating, varnishing, gumming, sludging, and coke formation.

3. **Dielectric strength** - A measure of resistance the oil has to electric current. Dissolved metals or moisture in the oil will lessen the resistance to electric current flow, thus giving the oil a lower dielectric strength. Oils of low dielectric strength may cause grounding of motor windings in hermetic compressors.

4. **Pour point** - The lowest temperature that the oil will flow at certain test conditions. The pour point should be well below the lowest temperature obtained in the evaporator. The amount of paraffin wax content in the oil determines the pour point. Higher pour points indicate greater wax content. If the pour point is reached in a low temperature application, the oil will congeal (curdle) in the evaporator causing low heat transfer and a loss in efficiency.

5. **Cloud point** - The temperature at which the wax will begin to precipitate out of the oil. The oil will become very cloudy at this temperature. The precipitated wax will plug metering devices and reduce evaporator heat transfer. The lower the cloud point, the better. This is a very important property for low temperature applications.

6. **Floc point** - The temperature at which wax will precipitate from a mixture of 10 percent oil and 90 percent refrigerant. The floc point is a more realistic point because most oils do mix with refrigerants. This test can be ignored when a nonmiscible refrigerant is employed. Very important for low temperature applications.

7. **Flash point** - The temperature of the oil when the oil is heated and a flame is passed over its surface causing a flash. The oil may momentarily flash, but may not continue to burn. The higher the flash point, the better.

8. **Fire point** - The temperature at which the oil will continue to burn once the flash point is reached. Higher fire points are favorable.

9. **Oxidation value** - Oil is heated in an atmosphere of oxygen and the sludge formed is weighed. This will predict the amount of sludge an oil may produce in an operating environment if conditions are right.

10. **Low temperature miscibility** - The temperature at which a mixture of oil and refrigerant will start to separate. An important property because refrigerant and oil must mix at low temperatures. Oil and refrigerant can separate (two-phase) in the evaporator. Two-phasing can cause oil hang-up in the evaporator because of the high viscosity of the oil when cold. Poor heat transfer and poor oil return to the compressor are the results. High temperature miscibility is also important because temperatures often reach 300° F in refrigeration systems.

11. **Foaming** - Oil foaming depends on the amount of refrigerant dissolved in the oil. Oil foaming is also a characteristic of the oil itself. Foaming usually occurs in the crankcase at start-up, but may occur during the running cycle. When a compressor is shut down or cycled off, refrigerant will migrate slowly to the oil in the crankcase because of a difference in vapor pressure. The oil will have the lowest vapor pressure at any temperature. An automatic pump down system

or crankcase heater will prevent this from happening. At the instant of start-up, the sudden pressure drop on the crankcase will cause refrigerant to come out of solution rapidly, thus foaming the oil. The oil-refrigerant foam may now generate excessive pressures in the crankcase and cause some of the oil to sneak past the compressor clearances and be pumped into the high side of the refrigeration system. The foaming oil can cause the compressor's moving parts to wear rapidly from loss of lubrication. Poor motor cooling will also result because of the crankcase being robbed of oil. Agitation of the oil by the crankshaft and its counterbalances can also cause oil foam. Foaming caused by agitation is usually much less severe.

12. **Material compatibility** - Soaking materials of construction in an oil and refrigerant mixture is a test procedure to check if the oil or refrigerant is compatible with the materials in a system. Elastomers, polyesters, copper, aluminum, etc., are found throughout the refrigeration system. The oil or refrigerant must not react, swell, shrink, deteriorate, weaken, pit, embrittle, or extrude any of these materials.

A service technician can now see what a science it would be if it were not for manufacturer's specifications and literature to assist him in using the proper oil for the application. Looking at all of the above properties is interesting, but can become very confusing when choosing an oil to use for a certain compressor. This is why it is recommended to let the compressor manufacturer specify what oil to use in any application. If you are unsure of what oil to use, call the compressor manufacturer or use their specification data along with their performance curve literature.

The compressor specification chart on page 53 is an example of compressor specification data. Note that the viscosity is 150 (SSU) for this compressor. Nothing is mentioned about additives even though they may be incorporated in the oils. The data also gives the amount of oil that the compressor was initially charged with (80 oz.) and the oil recharge

(72 oz.). Every conscientious service technician should have notebooks of every major compressor manufacturer's published performance curves and specifications data of their manufactured compressors. This information should be carried in the service vehicle at all times.

Oil Groups

Oils in general can be categorized into three basic groups. The groupings are animal, vegetable, and mineral oils (see table below). Both animal and vegetable oils cannot be refined or distilled without a change in composition. Both are considered poor lubricants in the refrigeration industry because of this changing composition. Poor stability is another disadvantage of animal and vegetable oils. These oils will form acids and gums very easily. Another problem with animal and vegetable oils in refrigeration applications is their somewhat fixed viscosity. Different viscosities for the diverse temperature applications in the refrigeration and air conditioning industries are mandatory and cannot be reached with animal or vegetable oils.

Three Basic Oil Groups
• Animal Oils
• Vegetable Oils
• Mineral Oils

Mineral Oils

The compounds in refrigeration mineral oils can be categorized into three main groups. They are paraffinic, naphthenic, and aromatics (see pie chart on page 54). The paraffinic group is refined from crude oil of the eastern United States region. The naphthenic group is refined from California and Texas crudes. Mid-continental crudes provide a mixture of both. It is the naphthenic group that constitutes most of the mineral oils used today in refrigeration because of their low wax contents and

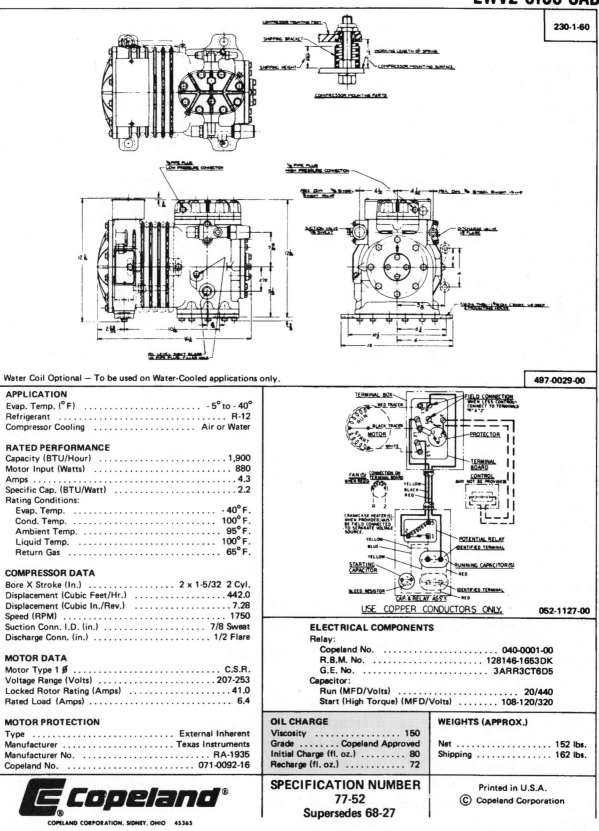

230-1-60

497-0029-00

Water Coil Optional — To be used on Water-Cooled applications only.

APPLICATION

Evap. Temp. (°F)	-5° to -40°
Refrigerant	R-12
Compressor Cooling	Air or Water

RATED PERFORMANCE

Capacity (BTU/Hour)	1,900
Motor Input (Watts)	880
Amps	4.3
Specific Cap. (BTU/Watt)	2.2

Rating Conditions:

Evap. Temp.	-40° F.
Cond. Temp.	100° F.
Ambient Temp.	95° F.
Liquid Temp.	100° F.
Return Gas	65° F.

COMPRESSOR DATA

Bore X Stroke (In.)	2 x 1-5/32 2 Cyl.
Displacement (Cubic Feet/Hr.)	442.0
Displacement (Cubic In./Rev.)	7.28
Speed (RPM)	1750
Suction Conn. I.D. (in.)	7/8 Sweat
Discharge Conn. (in.)	1/2 Flare

MOTOR DATA

Motor Type 1 Ø	C.S.R.
Voltage Range (Volts)	207-253
Locked Rotor Rating (Amps)	41.0
Rated Load (Amps)	6.4

MOTOR PROTECTION

Type	External Inherent
Manufacturer	Texas Instruments
Manufacturer No.	RA-1935
Copeland No.	071-0092-16

USE COPPER CONDUCTORS ONLY.

052-1127-00

ELECTRICAL COMPONENTS

Relay:

Copeland No.	040-0001-00
R.B.M. No.	128146-1653DK
G.E. No.	3ARR3CT6D5

Capacitor:

Run (MFD/Volts)	20/440
Start (High Torque) (MFD/Volts)	108-120/320

OIL CHARGE

Viscosity	150
Grade	Copeland Approved
Initial Charge (fl. oz.)	80
Recharge (fl. oz.)	72

WEIGHTS (APPROX.)

Net	152 lbs.
Shipping	162 lbs.

SPECIFICATION NUMBER
77-52
Supersedes 68-27

Printed in U.S.A.
© Copeland Corporation

low pour point values. Naphthenic based oils also have a lower viscosity index for the same temperature when compared with paraffinic based oils. Paraffinic based oils are recommended for electric motor lubrication. Paraffinic based oils have excellent chemical stability, but have somewhat poor solubility with polar refrigerants. Aromatics are a bit more reactive but have good boundary lubrication properties.

Synthetic Oils

Because of the somewhat limited solubility of mineral oils with certain refrigerants such as R-22, synthetic oils for refrigeration applications have been used successfully. With ozone depletion and global warming scares, much more research is being done with synthetic oils. Three of the most popular synthetic oils are alkylbenzenes, glycols, and ester based oils.

Three Popular Synthetic Oils
• Alkylbenzenes
• Glycols
• Esters

Alkylbenzene

Both naphthenic and paraffinic mineral oils are direct products of crude oil. However, alkylbenzene is synthesized from the raw materials propylene and benzene. Its synthetic origin distinguishes it from naphthenic and paraffinic mineral oils. One advantage of an alkylbenzene is that it is not dependent on the quality of any crude oil. Therefore, it will always have consistent compositions. Benzene and propylene can be bought at just about any refinery. The compound alkylbenzene is an aromatic hydrocarbon, and consists of only one type of compound (see pie chart next column). Hydrocarbons consist of molecules that contain hydrogen and carbon atoms. Most modern day refrigerants have

either ethane or methane as their base molecule. Ethane and methane are both hydrocarbon molecules. Aromatic hydrocarbons also make up mineral oils but in widely varying compositions.

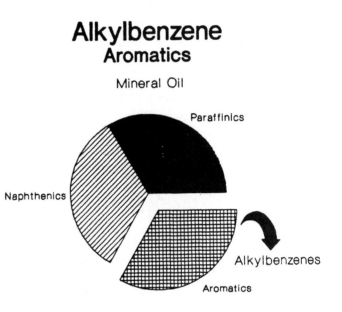

Alkylbenzene
Aromatics
Mineral Oil

Most HCFC-based refrigerant blends perform best with alkylbenzene lubricants when compared to other paraffinic and naphthenic mineral oils. This is because existing mineral oils are not completely soluble in the refrigerant blends. The blends are soluble in a mixture of mineral oil and alkylbenzene up to a 20 percent concentration of the mineral oil. This indicates that mineral oil systems that are retrofitted with refrigerant blends will not require extensive flushing of the oil. Today, as in the past, alkylbenzenes are used quite often in refrigeration applications (i.e., Zerol).

Glycols

Some of the most popular glycol-based lubricants are polyalkylene glycols (PAGs). They were the first generation lubricants used with HFC-134a. HFC-134a is a very polar molecule which contributes to its low solubility in non-polar lubricants such as mineral oils and older synthetic oils used in refrigeration applications. HFC-134a is a refrigerant that is replacing CFC-12 in medium and high temperature applications. The automobile industry has accepted HFC-134a because of its low hose permeability, along with satisfactory efficiencies. Because of the commercial availability of PAG oils, widespread testing is being done to evaluate the lubricant. However, many polyalkylene glycol lubricants tested with HFC-134a are not fully soluble and will separate. PAG oils also have a track record of being very hygroscopic (attract and retain moisture). However, modified PAGs are still being researched. Another disadvantage of PAG oils is their poor aluminum on steel lubricating abilities. PAGs also have been known to have a reverse solubility in refrigeration systems. This means that the oil may separate in the condenser instead of the evaporator. Today, many mineral oils separate in the evaporator. PAGs also have a very high molecular weight and can be harmful if inhaled in certain concentrations. One of the major drawbacks with PAGs are their incompatibility with chlorine. Because of this, the field retrofitting of CFC-12 systems with HFC-134a will be very difficult. HFC-134a systems will have to come from the factory as virgin systems with PAG oil if they are to be used in our industry.

Esters

Other synthetic oils gaining popularity are the ester based oils. A major advantage of ester based oils is their wax free composition. No wax will give a lower pour point. A popular ester-based oil is Polyol Ester. Polyol Esters have stable five-carbon neopentyl alcohols which, when mixed with fatty acids, will form the Polyol Ester family. They have been used as jet engine lubricants for years. Polyol Esters are far less hygroscopic than polyalkylene glycols. Some of these oils have very promising properties. Ester-based oils are the second generation oil for use with certain applications incorporating HFC-134a. Systems will have to come from the factory as virgin systems with ester oils for refrigeration use, or the technician will have to go through a step-by-step

flushing process to rid the system of most of its mineral oil. There are a few ester-based oils that are compatible with HFC-134a and a low percentage of mineral oil. However, a retrofit flushing process will have to be followed to rid the system of most of its mineral oils. Not more than 1 to 5 percent of the original mineral oil can be left in the system. Procedures have been established for a step-by-step flushing process to incorporate non-ozone depleting refrigerants and ester oils in active CFC systems incorporating mineral oils from 2 to 2000 hp. A new expansion valve and a liquid line drier should be the only equipment change involved in the retrofit. Ester-based oils are also used extensively in many HFC-based refrigerant blends. Ester-based lubricants are also used on many CFC and HCFC refrigerants and refrigerant blends. **Ester oils are very hygroscopic meaning they will attract and retain moisture easily.** However, always consult with the compressor manufacturer before using any lubricant.

Oil Additives

Oil additives fall into three general groups. They are polar compounds, polymers, and sulfur and chlorine compounds (see listing at top of the next page). Very satisfactory performance has been accomplished with oil additives. Additives can lower floc and pour points, improve thermal stability, cause antifoaming, inhibit oxidation, improve viscosities, decrease metal activation, prevent rusting, decrease metal wear, and handle extreme pressure situations. Oil additives will often help in one area, but be objectionable in others. All oil additives must be compatible with materials of construction. Oil additives can also be combined to give favorable properties. The service technician may not even know if additives are in the oil of the serviced compressor. This is why it is so important to consult with the compressor manufacturer or use the data specified by the manufacturer when adding oil.

Oil Additives
3 General Groups
• Polar compounds
• Polymers
• Sulfur and chlorine compounds

As service technicians, it is important to realize the magnitude of the refrigerant and oil transition this industry is experiencing. Refrigerants and oils have become a complex science. There used to be "rules of thumb" to follow that matched a certain viscosity with the temperature application. The diversification of oils and oil additives used with today's ozone friendly refrigerants and even yesterday's refrigerants make these rules of thumb obsolete. Education through reading current literature is one method a technician can use to keep abreast of these new technologies and changes in our industry. A technician can no longer rely on the rule of thumb for adding oil to a system. Technicians must always refer to the manufacturer's literature for each compressor to get information on what oil to incorporate.

Waste Oil

Although the Environmental Protection Agency (EPA) has recently ruled that refrigeration oils are not hazardous wastes, disposing of used oils in a careless fashion is against the law.

The EPA has specifically exempted "on condition that these used oils are not mixed with other wastes, that the used oils containing CFCs are subjected to recycling and/or reclamation for further use, and that these used oils are not mixed with used oils from other sources."

Used oil is hazardous (EP toxic) if a tested sample is found to contain specific compounds. Concentrations of specific compounds such as mercury, cadmium, or lead, or if the waste exhibits characteristics of ignitibility or corrosiveness, fall under description according to EPA for hazardous waste management.

Under current regulations, a used oil handler must determine (through testing or knowledge) that the used oil does not exceed the regulatory limits of the toxicity characteristic (TC) rule for constituents. Used oil that fails the TC must be disposed of according to hazardous waste regulations. Used oil that does not exceed the toxicity characteristic is not a hazardous waste.

The Agency has determined that properly drained used oil filters do not exhibit the toxicity characteristic. Therefore, it is not necessary to list used oil filters as a hazardous waste. EPA continues to encourage recycling of used oil from filters, and recycling of the filters and their components.

These rules were modified in 1986 to include small hazardous waste generators into the regulations (like service contractors who handle used oil and refrigerants).

Oil removed from refrigeration systems by your company could very well be classified as hazardous waste, if tested and found with concentrations of hazardous compounds. If your company disposes of 15 gallons (about 50 kg. or 110 pounds) of potentially hazardous waste, it could qualify as a small quantity generator.

It remains our responsibility to determine if our waste is hazardous. We are obligated to make sure that the waste, if hazardous, is disposed of safely and legally. Basically, it is our waste. We own it ... Forever.

The 1986 requirements are in three categories:

I. Generators of no more than 100 Kg./month

If you generate no more than 100 kg. (about 220 lbs or 25 gallons) of hazardous waste and no more than 1 kg. (about 2.2 lbs) of acutely hazardous waste in any calendar month, you are a conditionally exempt small quantity generator and the federal hazardous waste law requires you to:

1. Identify all hazardous waste you generate.

2. Send this waste to a hazardous waste facility, a landfill, or other facility approved by the state for industrial municipal wastes.

3. Never accumulate more than 1,000 kg. (about 2,200 lbs or under 300 gallons) of hazardous waste on your property. (If you do, you become subject to all requirements applicable to 100-1,000 kg./month generators.)

II. 100-1,000 kg./month generators (between 220 and 2,200 lbs or about 25 to under 300 gallons)

1. Comply with the 1986 rules for managing hazardous waste, including accumulation, treatment, storage, and disposal requirements.

2. Never accumulate more than 6,000 kg. (about 13,200 lbs or under 900 gallons) of hazardous waste on your property. If you do, you must comply with all applicable hazardous waste management rules.

III. Generators of 1,000 kg./month or more (about 2,200 lbs or 300 gallons or more) or 1 kg. of acutely hazardous waste in any month, the federal hazardous waste laws require you to:

1. Comply with all applicable hazardous waste management rules.

What's Your Category?

Add up the weight of all hazardous wastes your business generates (or collects) during a month. The total will determine your waste generator category.

Count the volume of used oil you generate; perhaps a small percentage of overall waste, but failure to account for it in the assessment could significantly impact your business.

Measure all quantities of hazardous waste that you:
- Accumulate on-site for any period of time prior to management.

- Package and transport off-site.

- Place directly in a regulated on-site treatment and disposal unit.

- Generate as still bottoms or sludges and remove from product storage tanks.

You do not have to measure wastes that:
- Are specifically exempt from the EPA.

- You recover continuously on-site without storing prior to reclamation.

- Are left as residue in the bottom of product storage tanks, if the residue is left in the tanks.

- Are left in the bottom of emptied containers.

- That you manage in an elementary neutralization unit, a totally enclosed treatment unit, or a waste water treatment unit.

- Are discharged directly to a publicly owned treatment works (POTW) without being stored or accumulated first. The POTW must also comply with the Clean Water Act.

- You have already measured once during the calendar month, and treated on-site or reclaimed in some manner, and used again.

If you decide to accumulate hazardous waste until you have enough to transport to a licensed hazardous waste management facility more economically, make sure:

1. You accumulate no more than 6,000 kg. of hazardous waste in any 180-day period if you are a 100 to 1,000 kg./month generator (or 270 days if you must transport your waste over 200 miles to a licensed hazardous waste facility).

2. You accumulate no more than 1,000 kg. of hazardous waste at any time if you are a generator of no more than 100 kg./month.

Used oils that contain CFCs after the CFC reclamation procedure, however, are subject to specification limits for used oil fuels if these oils are destined for burning.

Notes

SECTION FOUR
Ozone Depletion and Global Warming

- ❏ Ozone Depletion
- ❏ Radiation
- ❏ Stratospheric Ozone Depletion
- ❏ Global Warming
- ❏ Direct Effects
- ❏ Indirect effects
- ❏ Total Equivalent Warming Impact

Ozone Depletion

Ozone (O_3) is a gas that is found in both the stratosphere and troposphere, Figure 4-1. Ozone is a molecule that consists of three oxygen atoms (Figure 4-2) instead of the standard two atom oxygen or diatomic oxygen, (Figure 4-3). Stratospheric ozone is considered "good" ozone, because it shields the earth from harmful ultraviolet (UV-B) radiation. This three atom structure enables stratospheric ozone

to absorb harmful ultraviolet light from the sun. Tropospheric ozone is considered "bad" or unwanted ozone because it is a pollutant.

Stratospheric ozone, or "good" ozone, resides in the stratosphere which is between 7 and 30 miles above the earth's surface. This ozone accounts for over 90% of all ozone; however, it is rapidly being depleted by man-made chemicals containing chlorine, including refrigerants such as CFCs and HCFCs.

Tropospheric ozone, or "bad" ozone, resides in the troposhere, which extends from ground level to about 7 miles. This ozone accounts for about 10% of all ozone. The troposphere contains 90% of the atmosphere and is well mixed by weather patterns. Tropospheric ozone is formed by reactions between

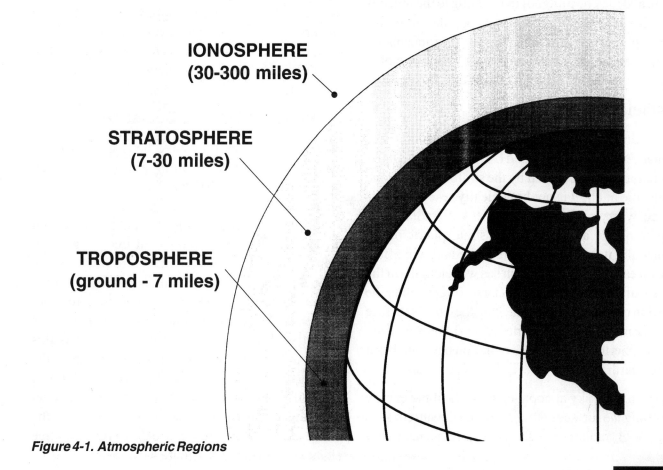

IONOSPHERE
(30-300 miles)

STRATOSPHERE
(7-30 miles)

TROPOSPHERE
(ground - 7 miles)

Figure 4-1. Atmospheric Regions

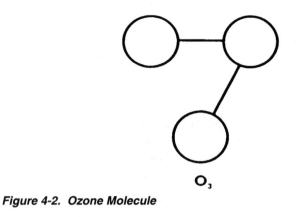

Figure 4-2. Ozone Molecule

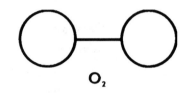

Figure 4-3. Diatomic Oxygen

hydrocarbons and oxides of nitrogen in sunlight. It has a **bluish color** when seen from the surface of the earth and may be **pungent** or irritating to the mucous membranes when inhaled. A popular term for tropospheric ozone pollution is smog. Tropospheric ozone contributes to the greenhouse effect, or global warming, which will be covered later in this section.

Radiation

The sun emits three types of ultraviolet (UV) radiation: UV-A, UV-B, and UV-C radiation. UV-A radiation is not absorbed by the stratospheric ozone molecule at all; it is not biologically reactive and is of no worry to life on earth. UV-B radiation is preferentially absorbed by the stratospheric ozone molecule; it is biologically reactive and will affect life on earth. UV-B radiation has wavelengths in the range of 280 to 320 manometers. It represents the portion of solar radiation reaching the earth's surface that is most efficiently filtered and controlled by stratospheric ozone. UV-C radiation never makes it to the earth.

Under a clear sky at noon, up to 0.5% of the energy reaching the surface of the earth from the sun consists of UV-B radiation. The intensity is primarily controlled by the angle of the sun, so that during winter or in the morning and evening, it is only a small fraction of that at noon in the summer. Clouds absorb and scatter a significant amount of UV-B radiation, as do trace gases (e.g., sulfur dioxide and nitrogen dioxide) and atmospheric particulates. A complicating factor is that ozone in the lower atmosphere, such as that associated with air pollution, can absorb UV-B that has passed through the stratospheric ozone layer.

Excess ultraviolet radiation causes the lens of the eye to cloud up with **cataracts**, which may cause blindness if left untreated. Ultraviolet radiation can also lead to **skin cancers**, including the often deadly melanoma. Increased exposure to ultraviolet radiation affects the body's immune system and ability to fight off disease. High doses of UV radiation also **reduce crop yields** worldwide. UV-B is the most dangerous of the ultraviolet radiations and can penetrate many meters below the surface of the ocean, killing marine life. This radiation has been known to kill photoplankton (one-celled plants) and krill (tiny shrimp), which are both at the bottom of the food chain in the ocean. Almost all ocean dwellers rely on a form of photoplankton and krill for survival.

Stratospheric Ozone Depletion

Oxygen molecules (O_2) have been generated for millions of years through a process called photosynthesis, which takes place in the oceans and on the earth. Oxygen molecules travel to the stratosphere where ultraviolet radiation from the sun breaks them apart into individual free oxygen (O) atoms. The free oxygen (O) atoms are very unstable and need to bond to other atoms or molecules. Some of these free oxygen (O) molecules bond to other oxygen (O_2) molecules, which is how ozone (O_3) is formed in the stratosphere, Figure 4-4. It is this ozone in the stratosphere that shields us from the sun's harmful UV-B radiation. Once ozone is formed, much of it is easily destroyed by the sun's ultraviolet radiation. The sun's radiation will break up the ozone molecule into diatomic oxygen (O_2) and free elemental oxygen (O), Figure 4-5.

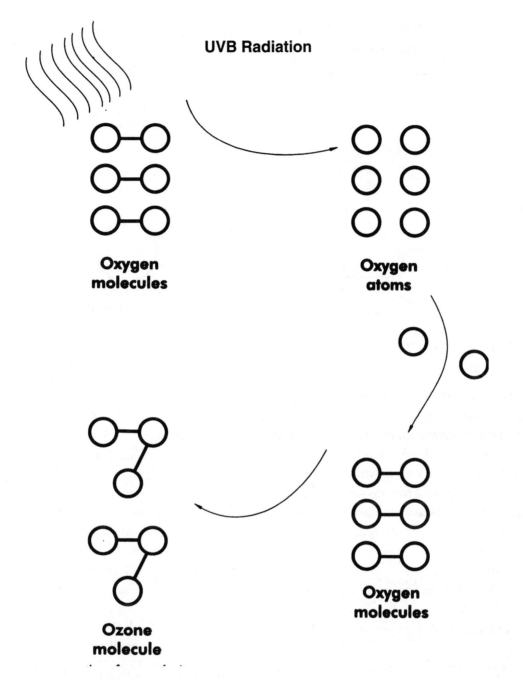

UVB Radiation

Oxygen molecules

Oxygen atoms

Oxygen molecules

Ozone molecule

Figure 4-4. The formation of stratospheric ozone

Ozone is constantly being created and destroyed in the stratosphere; the balance has been going on for millions of years. However, man-made chemicals, which are inert compounds, containing chlorine, such as CFCs and HCFCs, have knocked this delicate process out of balance in the last few decades. Volcanic eruptions and other man-made chemicals also contain chlorine, but their contributions are almost negligible when compared to the amount of chlorine emitted into the atmosphere by CFCs and HCFCs. The result is that stratospheric ozone is being depleted faster than it is being generated, and too much harmful ultraviolet radiation (UV-B) is reaching the earth.

CFC and a few HCFC molecules emitted into the troposphere reach the stratosphere by tropospheric winds. CFC molecules are very stable and have a

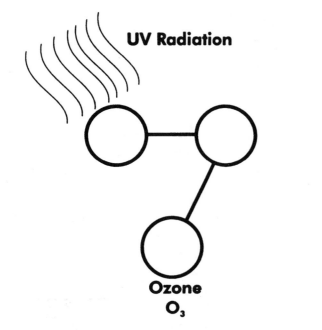
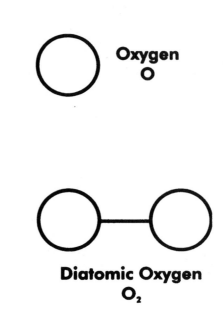

Figure 4-5. Breaking up an ozone molecule

long life that may extend over 100 years. In fact, if production and use of CFCs were to cease today, their effect on the ozone layer would not disappear until the year 2050 and beyond. CFCs neither dissolve in water nor break down into components that dissolve in water. **They do not "rain out" of the atmosphere.** HCFC molecules have much shorter lives and few ever reach the stratosphere.

Once in the stratosphere, ultraviolet radiation from the sun breaks off a chlorine atom (Cl) from the CFC or HCFC molecule. This chlorine atom is unstable and floats around looking for another molecule with which to bond. Upon finding an (O_3) molecule, the chlorine atom breaks it up into diatomic oxygen (O_2) and chlorine monoxide (ClO), Figure 4-6. This is the process by which ozone is depleted. An oxygen molecule is normally a diatomic molecule and wants to exist in pairs to be stable. A free oxygen (O) atom is very unstable and wants to combine with the unstable chlorine atom. The two molecules formed by this break-up are chlorine monoxide (ClO) and diatomic oxygen (O_2).

When the sun's ultraviolet radiation breaks diatomic oxygen (O_2) into unstable, individual free oxygen (O) atoms, these unstable oxygen atoms float around

the stratosphere looking to bond with another atom or molecule. Upon finding a chlorine monoxide (ClO) molecule, the free oxygen atom breaks off the chlorine atom from the molecule and bonds to the now free oxygen atom to form a stable diatomic molecule of oxygen (O_2). This reaction happens because diatomic oxygen (O_2) is more stable than chlorine monoxide (ClO). The free chlorine atom is left to float around the stratosphere looking for another ozone (O_3) molecule to attack and break up. This starts the ozone depletion process all over again. The chlorine atom is actually a catalyst in these ozone depleting reactions. It enters a reaction and comes out unchanged and unharmed, ready to start other reactions. It is believed that **one chlorine atom can destroy up to 100,000 ozone molecules,** which is why chlorine in the stratosphere is so destructive to the ozone layer.

It is extremely cold temperatures, sunlight, and clouds particular to the polar region where ozone is depleted. That's why ozone depletion is most pronounced over Antarctica. This is because of a polar vortex or swirling air mass, which has collected CFCs from industrialized nations over time. It is this swirling polar vortex that isolates the Antarctic

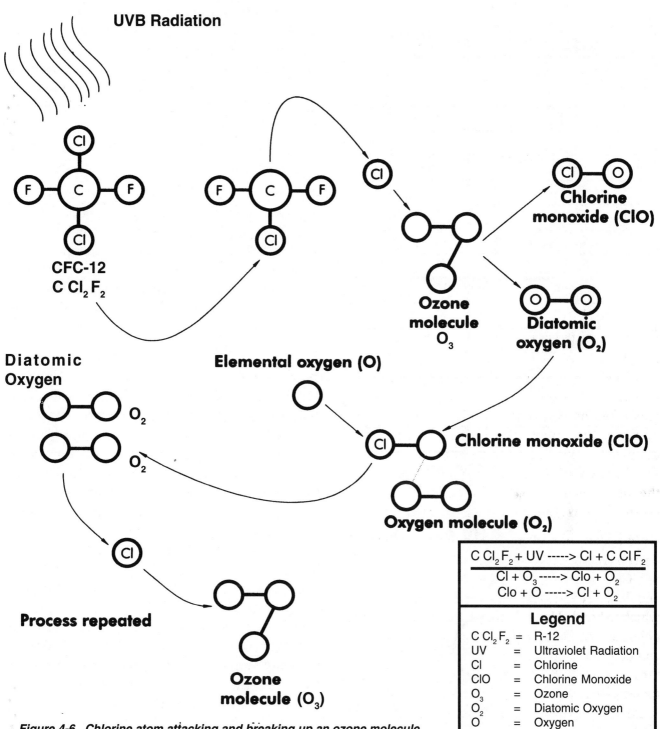

UVB Radiation

CFC-12
$C\ Cl_2\ F_2$

Chlorine monoxide (ClO)

Ozone molecule O_3

Diatomic oxygen (O_2)

Diatomic Oxygen O_2 O_2

Elemental oxygen (O)

Chlorine monoxide (ClO)

Oxygen molecule (O_2)

Process repeated

Ozone molecule (O_3)

$$C\ Cl_2\ F_2 + UV \dashrightarrow Cl + C\ Cl\ F_2$$
$$Cl + O_3 \dashrightarrow ClO + O_2$$
$$ClO + O \dashrightarrow Cl + O_2$$

Legend

$C\ Cl_2\ F_2$	=	R-12
UV	=	Ultraviolet Radiation
Cl	=	Chlorine
ClO	=	Chlorine Monoxide
O_3	=	Ozone
O_2	=	Diatomic Oxygen
O	=	Oxygen

Figure 4-6. Chlorine atom attacking and breaking up an ozone molecule

atmosphere from the rest of the world. The Antarctic region has very cold air, which causes tiny clouds of ice crystals to form in its stratosphere even though the humidity is low. When ultraviolet radiation breaks down the CFC molecules, chlorine monoxide (ClO) and other inorganic forms of chlorine cling to the ice crystals. Spring sunlight melts the ice crystals, releasing chlorine monoxide and other chlorine compounds to further react with ozone and deplete it in large quantities. These ozone-depleted masses

Refrigrant	ODP
R-11	1.0
R-12	0.93
R-114	0.71
R-115	0.38
R-22	0.055
R-123	0.016
R-125	0.0
R-134a	0.0
R-143a	0.0
R-152a	0.0
R-401A	0.037
R-410A	0.0
R-402A	0.019
R-402B	0.030
R-404A	0.0
R-407C	0.0

Table 4-1. Ozone depletion potentials

of air are then spread throughout the stratosphere by stratospheric winds, which dilute ozone-rich masses of air in other parts of the stratosphere. This causes ozone depletion in other parts of the stratosphere, **creating a global problem**. Data from NASA shows that the ozone layer over the northernmost parts of the United States, Europe, Russia, and Canada is depleted in the early spring as much as 35% in some years.

An index called the ozone depletion potential (ODP) has been adopted for regulatory purposes under the United Nations Environment Programme (UNEP) Montreal Protocol. The ODP of a compound is a measure of the ability of a chemical to destroy ozone molecules. The ODP shows relative effects of comparable emissions of various compounds. Table 4-1 shows some refrigerants with their ozone depletion potentials.

When stratospheric ozone intercepts ultraviolet light, heat is generated, Figure 4-7. This generated heat causes stratospheric winds, which are the main forces behind weather patterns on earth. By changing the amount of ozone, or even its distribution in the stratosphere, the temperature of the stratosphere can be affected, which can seriously affect weather on earth.

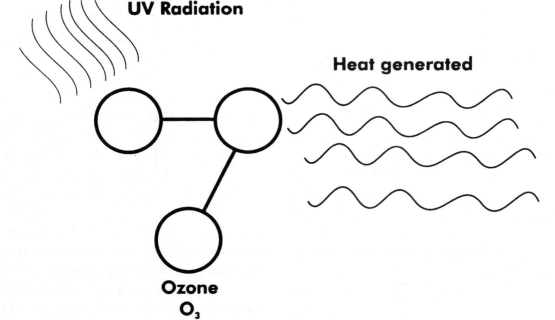

Figure 4-7. Heat generated from the break-up of an ozone molecule

Energy Waves

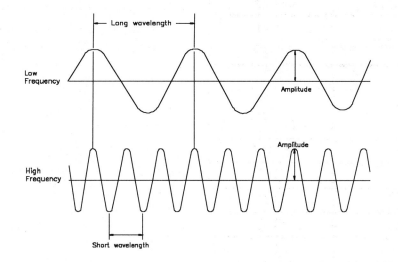

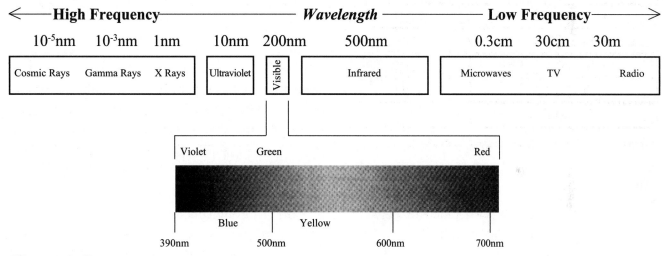

Figure 4-8. Energy waves

Global Warming

Global warming, or the greenhouse effect, is a completely different phenomenon than ozone depletion. Global warming refers to the physical phenomenon that may lead to the heating of the earth. While global warming and ozone depletion are two completely different issues, they can be caused by the same man-made chemicals such as refrigerants.

Most of the sun's energy reaches the earth as visible light. After passing through the atmosphere, part of this energy is absorbed by the earth's surface and is converted into heat energy. The earth, warmed by the sun, radiates heat energy back into the atmosphere toward space. Naturally occurring gases and lower atmospheric (tropospheric) pollutants such as CFCs, HCFCs, HFC, carbon dioxide, carbon monoxide, water vapor, and many other chemicals absorb, reflect, and/or refract the earth's infrared radiation and prevent it from escaping the lower atmosphere. This process slows the heat loss, making the earth's surface warmer than it would be if this heat energy had passed unobstructed through the atmosphere into space. The warmer earth surface then radiates more heat until a balance is established between incoming and outgoing energy. **This warming process, which is caused by the atmosphere's absorption of the heat energy radiated from the earth's surface, is called global warming** or the greenhouse effect. Increased concentrations of gases from man-made sources (e.g., carbon dioxide, methane, CFCs, HCFCs, and HFCs) that absorb the heat radiation cause global warming that could lead to a slow warming of the earth.

Global warming involves infrared radiation, which is less powerful and has a much lower frequency when compared to ultraviolet radiation, Figure 4-8. Re-radiated infrared radiation may cause a gradual increase in the average temperature of the earth. Over 70% of the earth's fresh water supply is either in ice cap or glacier form. Scientists are concerned that these ice caps or glaciers will melt if the average temperature rises too much, causing increased water levels. Global warming may also cause decreased crop yields and added smog levels on earth.

Considerable uncertainty exists about the climate change response to greenhouse gas emissions. This is due to the incomplete understanding of the following:

❑ Interactive feedback mechanism between clouds, oceans, and polar ice.

❑ Size and nature of the sources and sinks of the greenhouse gases.

Direct Effects

Chemicals that are emitted directly into the atmosphere and are not caused by any other related secondary process are direct emissions. These direct emissions are measured by an index referred to as the global warming potential (GWP). The GWP measures how much of a direct effect these emissions have on global warming. Refrigerant leaking from a refrigeration or air conditioning system is a good example of a direct emission, which has a direct effect on global warming. Refrigerants are measured

Refrigerant	GWP
R-11	1.0
R-502	4.1
R-12	3.0
R-22	0.34
R-123	0.002
R134A	0.28
R-401A	0.22
R-401B	0.24
R-404A	0.94
R-402A	0.51
R-402B	0.44
R-407C	0.34
R-410A	0.39

Table 4-2. Global warming potentials (direct effects)

compared to R-11, which has a global warming potential (GWP) of 1. Table 4-2 shows the GWPs for some popular refrigerants.

Carbon dioxide, CFCs, HCFCs, and HFCs are purged from the atmosphere at very different rates. Trees and plants are CO_2 scrubbers. Trees take in CO_2 and give off O_2 through photosynthesis. Carbon dioxide in the atmosphere decays very slowly. CFCs also decay very slowly, with rates varying from a 55-year lifetime for R-11 to a 550-year lifetime for R-115. Decay rates of HCFCs and HFCs also vary, but their lifetimes are much less than those of the CFCs, Table 4-3. The decay of an HFC compound with a mid-range lifetime is shown in Figure 4-9.

Indirect Effects

Because of its mass, quantity, long life, and chemical insulating effects, carbon dioxide (CO_2) is the number one contributor to global warming. Humanity has boosted levels of carbon dioxide in the atmosphere to over 25% the usual amount. Most of the CO_2 increase is caused by the combustion of fossil fuels. The combustion of fossil fuels, resulting in increased

Compound	Estimated Atmospheric Lifetime	GWPs for Various Integration Time Horizons		
		20 years	100 years	500 years
Carbon Dioxide	+	1	1	1
CFC-11	55	4500	3400	1400
CFC-12	116	7100	7100	4100
CFC-113	110	4600	4500	2600
CFC-114	220	6100	7000	5900
CFC-115	550	5500	7000	8500
HCFC-22	15.8	4200	1600	540
HCFC-123	1.7	330	90	30
HCFC-141b	10.8	1800	580	200
HCFC-225ca[1]	2.7	610	170	60
HCFC-225cb[1]	7.9	2400	690	240
HFC-32[2]	6.2	2440	720	260
HFC-125	40.5	5200	3400	1200
HFC-134a	15.6	3100	1200	400
HFC-143a	64.2	4700	3800	1600
HFC-152a	1.8	530	150	49
Methane[3]	10.5	35	11	4

+ The decay of carbon dioxide concentrations cannot be reproduced using a single exponential decay lifetime. Thus, there is no meaningful single value for the lifetime that can be compared directly with other values in this table.

[1] The HCFC-225ca/cb values were calculated by Atmospheric and Environmental Research, Inc. and are based on rate constant measurements reported by Z. Zhang et. al., Geophysical Research Letters, Vol. 18, January 1991, and the infrared energy absorption properties measured at AlliedSignal Central Research Laboratory.

[2] The HFC-32 values were calculated by Atmospheric and Environmental Research, Inc. and are based on the recommended value for the rate constant reported in Chemical Kinetics and Photochemical Data for Use in Stratospheric Modeling, Evaluation Number 10, JPL Publication 92-20, August 1992.

[3] The GWP values include the direct radiative effect and the effect due to carbon dioxide formation, but do not include any effects resulting from tropospheric ozone or stratospheric water formed as methane decomposes in the atmosphere.

Table 4-3. Integration time horizon (Courtesy, Alternative Fluorocarbons Environmental Acceptability Study)

CO_2, is what causes the indirect effects of global warming. Fossil fuel combustion is required for electricity, which is used to power refrigeration and air conditioning equipment. More efficient HVACR equipment requires less electrical energy, which means a decreased need for the combustion of fossil fuels and less CO_2 emitted into the atmosphere.

Indirect effects of global warming relate to the energy efficiency of the equipment. For example, refrigeration or air conditioning equipment that contains a relatively small charge of refrigerant that never leaks may still have a great impact on global warming. This is because the equipment may be undercharged or overcharged. The equipment is very inefficient under these conditions, and the carbon dioxide (CO_2) generated from longer run times contributes more to global warming than leaking refrigerant. This is an example of an indirect effect of global warming.

In the HVACR industry, scientists are mainly concerned with these indirect effects. Most of the newer refrigerants are more energy efficient than their predecessors; however, some are not as efficient. Just because newer refrigerants may not contribute to ozone depletion does not mean they do not contribute to the direct or indirect effects of global warming. An example of this is the newer refrigerant

R-134a. It has a zero ozone depletion index but does contribute to global warming directly and indirectly. Table 4-4 shows the indirect effects of refrigerants on global warming potential.

Total Equivalent Warming Impact

The total equivalent warming impact (TEWI) takes into consideration both the direct and indirect effects of refrigerants on global warming. Refrigerants with the lowest global warming and ozone depletion potentials have the lowest TEWI. Using HFC and HCFC refrigerants in the place of CFCs reduces the TEWI. All of the newer refrigerant alternatives introduced have a much lower TEWI.

Figure 4-10 shows that carbon dioxide from energy generation by a household refrigerator using HCFCs or HFCs would account for 96% of the contribution to global warming. If the HCFC or HFC refrigerant is recaptured at the end of the refrigerator's useful life, the direct contribution to global warming is

Refrigerant	GWP
CO_2	1
R-22	570
R-123	28
R-11	1300
R-12	3700
R-114	6400
R-115	13,800
R-134a	400

Table 4-4. Indirect effects of refrigerants on global warming potentials

eliminated. For refrigeration applications, the energy factor is so dominant that the difference between HCFC and HFC alternatives is insignificant, as long as the refrigerator efficiency is not compromised over its useful life. Figure 4-11 shows another example of TEWI.

Figure 4-12 compares baseline CFC refrigerants to HCFC/HFC alternatives and HCFC/HFC alternatives with deducted losses when applied to commercial chillers. TEWI consists of direct and indirect effects, fluorocarbon emission levels, global warming potentials, and carbon dioxide emissions coming from the equipment.

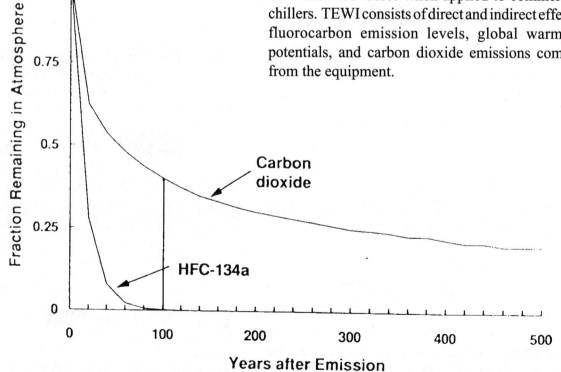

Figure 4-9. *Carbon dioxide and R-134a fractions in the atmosphere as a function of time (courtesy, Alternative Fluorocarbons Environmental Acceptability Study).*

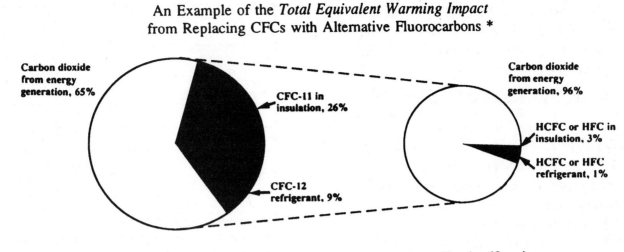

An Example of the *Total Equivalent Warming Impact*
from Replacing CFCs with Alternative Fluorocarbons *

Carbon dioxide
from energy
generation, 65%

CFC-11 in
insulation, 26%

CFC-12
refrigerant, 9%

Carbon dioxide
from energy
generation, 96%

HCFC or HFC in
insulation, 3%

HCFC or HFC
refrigerant, 1%

Total: 100 units

Total: 68 units

Relative contributions to future global warming of three gases emitted to the atmosphere as a result of operating a refrigerator over a 15-year period beginning in 1990.

Relative contributions to future global warming of three gases emitted to the atmosphere as a result of operating a refrigerator over a 15-year period beginning in 2000.

* Based on identical energy efficiency of the systems compared and a 100-year integration time horizon.

Figure 4-11. *An example of TEWI (Courtesy, Alternative Fluorocarbons Environmental Acceptability Study)*

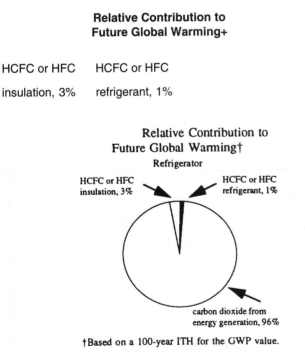

**Relative Contribution to
Future Global Warming+**

HCFC or HFC   HCFC or HFC

insulation, 3%   refrigerant, 1%

Relative Contribution to
Future Global Warming†

Refrigerator

HCFC or HFC
insulation, 3%

HCFC or HFC
refrigerant, 1%

carbon dioxide from
energy generation, 96%

†Based on a 100-year ITH for the GWP value.

Figure 4-10. *Relative contribution to future gloBal warming (Courtesy, Alternative Fluorocarbons Environmental Acceptability Study).*

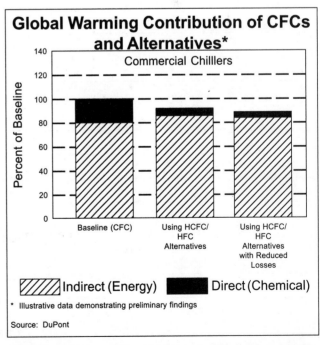

Global Warming Contribution of CFCs and Alternatives*

Commercial Chilllers

Percent of Baseline

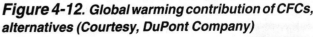

Indirect (Energy)   Direct (Chemical)

* Illustrative data demonstrating preliminary findings

Source: DuPont

Figure 4-12. *Global warming contribution of CFCs, alternatives (Courtesy, DuPont Company)*

Notes

SECTION FIVE
Montreal Protocol

❏ Early Concerns and Controls
❏ Montreal Protocol
❏ Chemicals (CFCs) included in the Montreal Protocol
❏ Reassessments and Updates to the Protocol

Early Concerns and Regulations

The 1974 Molina-Rowland theory that man-made CFCs and bromine emitted into the atmosphere are responsible for ozone depletion caused considerable debate and controversy. As a result of this theory, subsequent evidence, and wide spread public concern, the United States banned the use of CFCs in non-essential aerosols in 1978.

Montreal Protocol

In September 1987, the United States and 22 other countries signed the Montreal Protocol to control production of ozone depleting substances that include chlorofluorocarbons, hydrochlorofluorocarbons and halons. The Protocol contains provisions for countries not originally signing, and eventually bans imports and exports of regulated CFCs and products containing CFCs from non-signatory nations.

Chemicals (CFCs) Included in the Montreal Protocol

GROUP I - Fully-Halogenated Chlorofluorocarbons
    CFC-11
    CFC-12
    CFC-113
    CFC-114
    CFC-115

GROUP II - Halons
    Halon-1211
    Halon-1301
    Halon-2402

NOTE: Halons are chemicals used in industrial and commercial applications. Computer room fire suppression systems use halons to suffocate combustion. Halons are extremely high ozone depleting chemicals.

The Protocol froze production of Group I refrigerants back to 1986 levels starting in 1989, then also reduced production of each group by 20 percent in 1993, followed by an additional 30 percent in 1997 (see bar chart on next page).

Ozone Depletion Potential (ODP)		
Compound	**ODP**	**GWP** (global warming potential)
CFC-11	1.0 (Base)	1.0 (Base)
CFC-12	0.93	3.0
CFC-114	0.71	7.2
HCFC-22	0.055	0.34
HCFC-123	0.016	0.002
HFC-134a	0.0	0.28
R-401A	0.037	0.22
R-401B	0.039	0.24
R-402A	0.019	0.51
R-402B	0.03	0.44
R-404A	0.0	0.94
R-407C	0.0	0.34
R-410A	0.0	0.39
R-408A	0.024	0.64

Note: These ODPs have been assigned relative to the ability of these substances to deplete stratospheric ozone.

Reassessments and Updates to the Protocol

Since the research and assessment of ozone depletion is continuous, provisions for scheduled meetings and updating of the protocol were outlined. The first occurred in June of 1990, in London, where 56 nations signed an agreement that strengthened the 1987 agreement as follows:

Control of CFCs
(1987 Protocol)

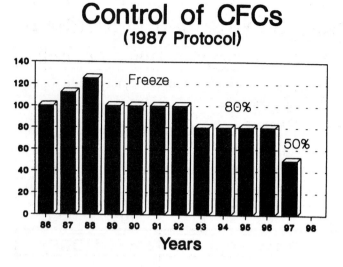

❑ Accelerated phaseout of CFCs, with complete phaseout by 2000.

Note: Canada and some of the European nations signed a separate agreement that would accelerate the phaseout to a 1997 total phaseout. President Bush ordered a 1995 phaseout for the U.S.

❑ Reduced production levels of CFCs (from 1986 levels, and the 1987 protocol) to a 50 percent reduction in 1995, an 85 percent reduction in 1997, followed by 100 percent in the year 2000, with each of these taking place on January 1. Germany also pledged to stop production by the end of 1996.

❑ Methyl chloroform to be reduced 30 percent in 1993, 70 percent in 2000, and 100 percent by 2005.

The November 1992 Protocol meeting in Copenhagen, Denmark resulted in the first-ever phaseout schedule of HCFCs and accelerated the Protocol phaseout of CFCs. The United States representation consisted of government, industry, and associations who collectively proposed and agreed with the outcomes of the meeting that will allow sufficient time for industry to develop long-term alternatives to HCFCs. This agreement was signed by 93 nations and resulted in the following:

❑ Parties are encouraged to recover, recycle, and reclaim controlled substances.

❑ The multilateral fund for technology transfer from developed to developing countries was made permanent with an approved budget through 1994.

❑ Methyl bromide consumption is frozen in 1995 to 1991 levels. Methyl bromide is a widely used agricultural chemical.

❑ Eliminates the use of CFCs as of January 1, 1996, except for essential uses, which must be agreed upon by the parties of the protocol.

❑ HCFC consumption will be limited or capped at a percent of historic usage beginning in 1996. The cap will incrementally decrease and ultimately eliminate world wide consumption of HCFCs in the year 2030. See phaseout schedules and impact charts that follow.

EPA, industry, and associations estimate that this phaseout will allow sufficient time for development and transition to long-term alternatives to HCFCs. Should annual production levels exceed the cap, the EPA has the authority to limit HCFC production

Phaseout Schedule Resulting from Copenhagen Agreements

Percentage of Base Year Production and Consumption (1992 Protocol)

Year	Compounds							
	CFCs 11, 12, 113, 114, 115	Halons 1301, 1211, 2402	CFC (Other) 13	Carbon Tetrachlor	Methyl Chloroform	HCFCs	HBFCs***	Methyl Bromide
Base Year	1986	1986	1989	1989	1989	1989	---	1991
1992	100%	100%	100%	N.C.	N.C.	N.C.	N.C.	N.C.
1993	100%	100%	80%		100%			
1994	25%	0%	25%		50%			
1995	25%		25%	15%	50%			Freeze
1996	0%		0%	0%	0%	Freeze at cap*	0%	
2000						100%		75%**
2004						65%		
2010						35%		
2015						10%		
2020						0.5%		
2030						0%		

N.C. - No Controls - Only Listed

* HCFC cap: (3.1% of ODP weighted CFC consumption in 1989) + (HCFC consumption in 1989)

**Suggested 25% cut to occur in 2000, decision to be based on subsequent assessments

***Specific list of HCFCs, HBFCs (Hydrobromoflourocarbons) listed in Annex C to Copenhagen Amendments

HCFC ODP Cap Impact
Million Kgs. CFC-11 Equivalent

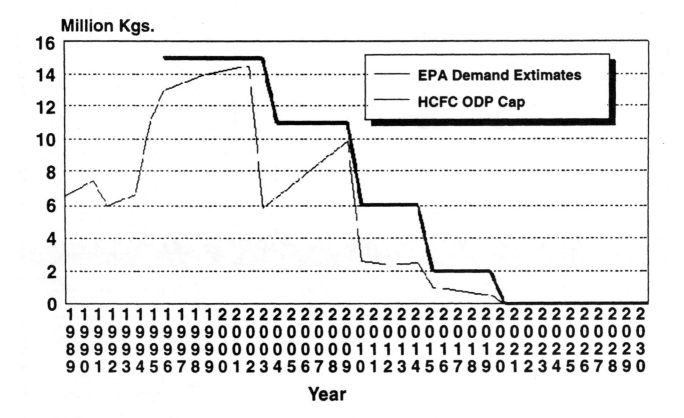

Million Kgs.

Legend:
- EPA Demand Extimates
- HCFC ODP Cap

Year

HCFC Phaseout Schedule

January 1, 1996 3.1 percent of 1989 CFC production (ODP weighted) combined with ODP weighted HCFC production in 1989 = **cap**.

NOTE: As worldwide production levels are gathered, the cap of each HCFC will be established.

January 1, 2004	65% of cap	No production and no importing of HCFC-141b
January 1, 2010	35% of cap	No production and no importing of HCFC-142b and HCFC-22, except for use in equipment manufactured before 1/1/2010 (so no production or importing for NEW equipment that uses these refrigerants)
January 1, 2015	10% of cap	No production and no importing of any HCFCs, except for use as refrigerants in equipment manufactured before 1/1/2020
January 1, 2020	0.5% of cap	No production and no importing of HCFC-142b and HCFC-22
January 1, 2030	0% of cap	No production and no importing of any HCFCs

Notes

Notes

SECTION SIX
Legislation and Regulation

- ❑ Clean Air Act Amendments
- ❑ Class I Substances
- ❑ Class II Substances
- ❑ Phaseout of Production and Consumption of Class I and Class II Substances
- ❑ National Recycling and Emission Reduction Program
- ❑ EPA Regulations
- ❑ Record Keeping
- ❑ Enforcement
- ❑ Recommended Forms

Clean Air Act Amendments

On November 15, 1990, President Bush signed the 1990 Clean Air Act Amendments (P.L. 101-549). Title VI establishes national policy for the reduction and ultimate elimination of stratospheric ozone depleting substances. The Act also addresses rules that will foster similar actions internationally.

The objectives of Title VI (Section 608) are to reduce the production, use, and emissions of ozone depleting substances to the lowest achievable level and promote the recapture and recycling of these substances. The substances affected are divided into classifications, as follows:

Class I Substances:

Group I:
    chlorofluorocarbon-11 (CFC-11)
    chlorofluorocarbon-12 (CFC-12)
    chlorofluorocarbon-113 (CFC-113)
    chlorofluorocarbon-114 (CFC-114)
    chlorofluorocarbon-115 (CFC-115)

Group II:
    halon-1211
    halon-1301
    halon-2402

Group III:
    chlorofluorocarbon-13 (CFC-13)
    chlorofluorocarbon-111 (CFC-111)
    chlorofluorocarbon-112 (CFC-112)
    chlorofluorocarbon-211 (CFC-211)
    chlorofluorocarbon-212 (CFC-212)
    chlorofluorocarbon-213 (CFC-213)
    chlorofluorocarbon-214 (CFC-214)
    chlorofluorocarbon-215 (CFC-215)
    chlorofluorocarbon-216 (CFC-216)
    chlorofluorocarbon-217 (CFC-217)

Group IV:
    carbon tetrachloride

Group V:
    methyl chloroform

Class II Substances:
    hydrochlorofluorocarbon-21 (HCFC-21)
    hydrochlorofluorocarbon-22 (HCFC-22)
    hydrochlorofluorocarbon-31 (HCFC-31)
    hydrochlorofluorocarbon-121 (HCFC-121)
    hydrochlorofluorocarbon-122 (HCFC-122)
    hydrochlorofluorocarbon-123 (HCFC-123)
    hydrochlorofluorocarbon-124 (HCFC-124)
    hydrochlorofluorocarbon-131 (HCFC-131)
    hydrochlorofluorocarbon-132 (HCFC-132)
    hydrochlorofluorocarbon-133 (HCFC-133)
    hydrochlorofluorocarbon-141 (HCFC-141)
    hydrochlorofluorocarbon-142 (HCFC-142)
    hydrochlorofluorocarbon-221 (HCFC-221)
    hydrochlorofluorocarbon-222 (HCFC-222)
    hydrochlorofluorocarbon-223 (HCFC-223)
    hydrochlorofluorocarbon-224 (HCFC-224)
    hydrochlorofluorocarbon-225 (HCFC-225)
    hydrochlorofluorocarbon-226 (HCFC-226)
    hydrochlorofluorocarbon-231 (HCFC-231)
    hydrochlorofluorocarbon-232 (HCFC-232)
    hydrochlorofluorocarbon-233 (HCFC-233)
    hydrochlorofluorocarbon-234 (HCFC-234)
    hydrochlorofluorocarbon-235 (HCFC-235)

hydrochlorofluorocarbon-241 (HCFC-241)
hydrochlorofluorocarbon-242 (HCFC-242)
hydrochlorofluorocarbon-243 (HCFC-243)
hydrochlorofluorocarbon-244 (HCFC-244)
hydrochlorofluorocarbon-251 (HCFC-251)
hydrochlorofluorocarbon-252 (HCFC-252)
hydrochlorofluorocarbon-253 (HCFC-253)
hydrochlorofluorocarbon-261 (HCFC-261)
hydrochlorofluorocarbon-262 (HCFC-262)
hydrochlorofluorocarbon-271 (HCFC-271)

The Act addresses global concerns by prohibiting the export of technology or the production of all Class I substances. It also provides technical and financial assistance to developing nations (that are signatories of the Montreal Protocol) to foster cooperative research, studies, and policy of a global scope.

Phaseout of Production and Consumption of Class I Substances

The Act originally phased out CFCs by the year 2000, with reductions each year until production ends, but was accelerated to phase out CFCs by December 31, 1995. (See CFC phase out chart below.)

Phaseout of Production and Consumption of Class II Substances

Currently, the EPA has established the following phaseout of HCFCs (see EPA Phaseout Schedule for HCFCs at top of next column). With this schedule,

which is subject to change, R-22 will be phased out in 2020 and will not be used in the production of new air conditioning equipment after January 1, 2010. The EPA has the authority to accelerate the phaseout of these substances if the U.S. consumption exceeds the HCFC cap as agreed upon in the Montreal Protocol. An accelerated phaseout of R-22 is thought to be extremely unlikely however.

EPA Phaseout Schedule for HCFCs		
Year	Substance	Phaseout*
2003	HCFC-141b	Production and consumption (used in foam production applications)
2010	HCFC-22, -142b	Production and consumption, except for use in equipment manufactured before January 1, 2020.
2015	Other HCFCs	Production and consumption, except for use in equipment manufactured before January 1, 2020.
2020	HCFC-22, -142b	No production and no importing.
2030	Other HCFCs	No production and no importing of any HCFCs.

*Does not apply to feedstock.

National Recycling and Emission Reduction Program (Section 608)

This is the key provision for HVACR contractors and technicians. Entitled the National Recycling and Emission Reduction Program, Section 608 seeks to reduce the use and emission of CFCs and HCFCs to the lowest achievable level and to maximize recapture and recycling.

Effective July 1, 1992, it became unlawful to knowingly vent, release, or dispose of CFCs and HCFCs during the repair, service, maintenance, or disposal of appliances or industrial process refrigeration equipment.

The venting prohibition for alternate (substitute) refrigerants such as HFCs became effective **November 15, 1995**. The term "refrigerant" will

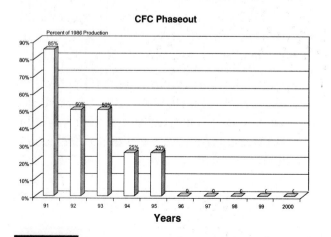

CFC Phaseout

Percent of 1986 Production

Years

include alternatives and be subject to the same regulations of not being vented.

NOTE: Only **three types of releases are permitted** under the prohibition:

1. **De minimus** is defined as the minimum amount of refrigerant that cannot be contained during a service practice.

2. **Refrigerants emitted in the course of normal operation** of air conditioning and refrigeration equipment, such as chiller purgers, as opposed to DeMinimus. (EPA is requiring the repair of substantial leaks, however.)

3. **Mixtures of nitrogen and R-22 that are used** as holding charges or **as leak test gases**, because as in these cases, the ozone-depleting compound R-22 is not used as a refrigerant.

EPA Regulations

Under Section 608 of the Clean Air Act, the EPA has issued regulations establishing five main requirements for recycling, emissions reduction, and disposal of Class I and Class II substances.

1. Service Practices

Requires technicians servicing and disposing of air conditioning and refrigeration equipment to observe service practices that reduce refrigerant emissions.

❑ The most fundamental of these practices is the requirement to recover refrigerant rather than vent it to the atmosphere. Venting that takes place with the knowledge of the technician during the maintenance, servicing, repairing, or disposal of air conditioning and refrigeration equipment, less de minimis, good faith releases, are in violation of the Act.

❑ Before air conditioning or refrigeration equipment is opened for maintenance, service, or repair, the refrigerant in either the entire system or the part to be serviced (if the latter can be isolated) must be transferred to a system receiver or to (by) a certified recycling or recovery machine.

2. Technician Certification

Requires technicians servicing air conditioning and refrigeration equipment to obtain certification through an EPA-approved testing organization and restricts sales of refrigerant to certified technicians.

❑ Technician is defined by the EPA as any person who performs maintenance, service, or repair to air conditioning or refrigeration equipment that could reasonably be expected to release CFCs or HCFCs into the atmosphere.

Technician also means any person who performs disposal of appliances (except for small appliances and motor vehicle air conditioning) that could reasonably be expected to release CFC or HCFC refrigerant into the atmosphere.

Activities **included** in the definition of technician are as follows:

❑ Attaching and detaching hoses and gauges to and from the appliance to measure pressure with the appliance.

❑ Adding or removing refrigerant to or from the appliance.

❑ Any other activity that violates the integrity of the refrigerant circuit while there is refrigerant in the appliance.

❑ Assembling split systems.

Activities **excluded** from the defintion of technician are as follows:

❑ Activities that are not reasonably expected to violate the integrity of the refrigerant circuit, such as painting, rewiring, replacing insulation, or tightening nuts or bolts on the appliance.

❑ Maintenance, service, repair, or disposal of appliances that have been evacuated pursuant to the EPA recovery/recycling regulations, unless the maintenance consists of adding refrigerant to the appliance.

❑ Servicing motor vehicle air conditioners, which are subject to EPA 609 regulations.

❑ Disposing of motor vehicle air conditioners (MVAC), MVAC-like appliances, and small appliances.

❑ Only technicians in possession of an EPA-approved certification identification card will be able to purchase Class I and Class II refrigerants, except for "small cans" that contain less than 20 pounds, in which case 609 certification is required.

❑ Only certified technicians may purchase bulk containers of 20 lbs. or more of refrigerant (cylinders and drums), operating system pre-charged parts, and pre-charged split systems containing Class I and II refrigerants. Fully assembled (manufactured) units—household refrigerators, window and packaged air conditioners, and systems containing HCFCs—are exempt from certification requirements relative to purchase and installation.

❑ Pre-charged split systems can be purchased only by certified technician (Type II). Pre-charged split systems are considered by the EPA to be a set of parts, at least one of which is pre-charged, from which one can assemble a complete system. This may include a pre-charged condenser, pre-charged evaporator, and pre-charged line set, or simply a pre-charged condensing unit sold with an evaporator and line set containing nitrogen. Note: The EPA has put a "stay" (hold) on this regulation pending a retailers objection.

❑ Refrigerant changing hands between manufacturer, wholesaler, etc., are exempt from the certification rule.

❑ Contractors are allowed to have "clerks" or "runners" (rather than technicians) pick up or purchase refrigerant by making prior arrangements with wholesalers and placing on file a photocopy of the technician certification card. It becomes the contractor's (individual purchasers or large purchasers) responsibility to notify wholesalers of changes in the status of certified employees.

❑ Types of technician certification
Persons successfully completing a core of questions on stratospheric ozone protection (25 each) and questions from one or more of the following types at 70 percent success, will be certified in those areas. Persons certified in all three areas will be "universally" certified. At this time, the EPA does not intend to require recertification. Staying abreast with future changes in regulation and technology is the responsibility of the technician.

Type I
Small Appliance
Manufactured, charged, and **hermetically sealed** with **five (5) pounds or less of refrigerant**. Includes refrigerators, freezers, room air conditioners, package terminal air conditioners ("PTACs"), heat pumps, dehumidifiers, under-the-counter ice makers, vending machines, and drinking water coolers.

Type II
High Pressure Appliance
Uses refrigerant with a boiling point between -50°C (-58°F) and 10°C (50°F) at atmospheric pressure. Includes 12, 22, 114, 500, and 502 refrigerants.

Very High Pressure Appliance
Uses refrigerants with a boiling point below -50°C (-58°F) at atmospheric pressure. Includes 13 and 503 refrigerants.

Type III
Low Pressure Appliance
Uses refrigerant with a boiling point above 10°C (50°F) at atmospheric pressure. Includes 11, 113, and 123 refrigerants.

Universal Certification
Certified in all the above: Type I, II, and III.

3. Equipment and Reclaimer Certification
Establishes equipment and reclaimer certification programs that would verify all recycling or recovery equipment sold was capable of minimizing emissions,

and that reclaimed refrigerant on the market was of known and acceptable quality to avoid equipment failures from contaminated refrigerants.

❑ Before systems are opened for maintenance, service, or repair, the technician must evacuate the refrigerant in either the entire unit or the part to be serviced (if the latter can be isolated) to a system receiver or to (by) a certified recovery or recycling machine. The vacuum levels are found in the tables of Section Eight.

❑ Contractors must certify that they own certified recovery or recycling equipment to perform on-site recovery or recycling.

❒ With the adoption of the Industry Regulation Guideline (IRG 2), refrigerant can change ownership as long as it is brought to ARI Standard 700 purity levels, and as long as the refrigerant remains in the contractor's custody and control. These purity levels can be accomplished through field recycling, testing, and documentation or by returning to a certified reclaimer. The following

four options are available for technicians/contractors with refrigerant that has been recovered from a system. See chart previous column.

Option 1: Put refrigerant back into the system without recycling it.

Option 2: Recycle refrigerant and put it back into the system from which it was removed or back into a system with the same owner.

Option 3: Recycle the refrigerant, test to verify conformance to ARI Standard 700 prior to reuse in a different owner's equipment provided that the refrigerant remains in the contractor's custody and control at all times from recovery through recycling to reuse.

Option 4: Send refrigerant to a certified reclaimer.

Leak Rate Chart			
50 pounds and _less_ of Refrigerant charge			Repair not required
50 pounds and _more_ of Refrigerant charge	Industrial process and commercial refrigeration equipment	35% Annual leak rate	Repair required
	Comfort cooling chillers and <u>ALL</u> other equipment	15% Annual leak rate	Repair required

4. Leak Repair of Systems Containing More than 50 pounds

Requires the repair of substantial leaks based on annual leak rates of systems containing **more than 50 pounds of refrigerant charge**.

❑ Leak repair will be required by owners of industrial process and commercial refrigeration equipment with an **annual leak rate of 35 percent**. For comfort cooling chillers and all other equipment, leak repair will be required for equipment with an **annual leak rate of 15 percent**.

IRG 2
Refrigerant Flowchart

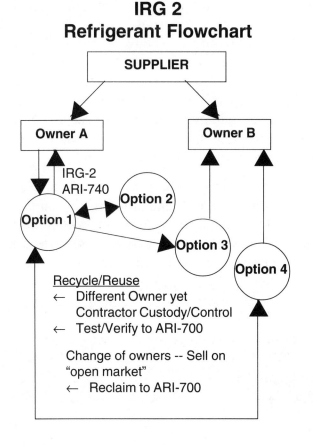

Owners of this equipment are responsible for the recordkeeping of refrigerant added to their equipment and refrigerant purchased.

See leak rate chart on previous page.

5. *Safe Disposal of Equipment and Appliances*

Requires that ozone depleting refrigerants in appliances, machines, and other goods be removed and recovered prior to their disposal or dismantling (e.g., commercial refrigeration, cold storage, chillers, and industrial process refrigeration equipment).

❑ Require that refrigeration equipment be provided with service apertures that would facilitate recovery. Small appliances will require a process stub for access.

❑ The final person in the disposal chain (e.g., a scrap metal recycler or landfill owner) is responsible for ensuring that refrigerant is recovered or for obtaining verification that the refrigerant has been removed previously.

Record Keeping

In addition to the five major elements established, the EPA has established the following record-keeping requirements to assist with enforcement of the Act:

❑ Equipment certification (recycle, recovery). The EPA is requiring the maintenance of equipment testing and performance records as well as a list of all equipment that meets EPA requirements.

❑ Technicians servicing equipment that contains 50 or more pounds of refrigerant must provide the owner with an invoice that indicates the amount of refrigerant added to the appliance.

❑ Reclaimers are required to maintain records of names, addresses, and quantity of refrigerant sent to them for reclamation.

❑ Contractors and other recovery and recycling equipment owners must certify that they own and are properly using certified recovery or recycling equipment.

❑ Disposers (persons disposing of small appliances) must maintain copies of signed statements that verify the refrigerant was recovered.

❑ Approved technician certification programs must maintain records of the names, addresses, and scores of persons certified along with the date and location of certification test. Certifying programs must send the EPA an activity report every six months that includes pass/fail rate and testing schedules.

❑ Wholesalers who sell CFC and HCFC refrigerant must retain invoices that indicate the name of the purchaser, the date of sale, and the quantity of refrigerant purchased.

❑ Owners/operators of air conditioning and refrigeration equipment must maintain records of servicing, service calls, and amounts of refrigerants added. Owners adding their own refrigerant shall keep records of refrigerant purchased and added to equipment each month.

Enforcement (Title VII)

The Act has a number of civil and criminal penalties that, upon conviction, would impose heavy fines and/or imprisonment. The regulations will affect violators at the administrative, manufacturing, distribution, and user levels. Violations include failing to report, falsifying documents, tampering with monitoring devices, **knowingly releasing** these substances into the ambient air, and using or manufacturing unapproved equipment.

The EPA would be authorized to seek legal redress through the local, state, and U. S. District Courts that would allow injunctions, fines, and imprisonment under Title 18 of the United States Code. The fines received from these infractions would be deposited into a "special fund in the United States Treasury for licensing and other services," and expended by the administrator.

The administrator of the agency (with the attorney general) will have the authority to process and initiate actions for violations which could result in the following:

- Civil **penalties of up to $27,500 per day per violation** of the Act.

- Revocation of certification, such as for not being able to demonstrate proper recovery procedures.

- For the encouragement of others to report violations, **awards up to $10,000** to persons furnishing information that leads to the conviction of a person violating provisions of the Act.

Stages of Enforcement

- SECTION 114 LETTER: A letter from the EPA to a suspected violator of the Act (contractor, building owner, technician, etc.) that charges a violation and requires additional information to the EPA from the accused party.

- FINDING OF VIOLATION: Notification of a violation letter from the EPA that documents a violation and places it on file.

- ADMINISTRATIVE ORDER: A stronger communication from the EPA directing compliance of the Act or closing down business.

- FIELD CITATION: The EPA may issue tickets for minor violations under authority of the Clean Air Act.

- ADMINISTRATIVE HEARING: A hearing called by the EPA for pursuing a violation less than $200,000 and less than 1 year old.

- CIVIL/JUDICIAL COURTS: Significant violations of the Clean Air Act that will be heard by the federal court.

Recommended Forms

While the Act does not require recordkeeping for systems containing less than 50 pounds of refrigerant, enforcement of the Act does require proof. The only way to defend accusations of venting is through a recordkeeping system that tracks all refrigerant purchased, recovered, sent for reclamation, and sold to customers at each job site.

For a contractor to account for refrigerant, the technician must be responsible for refrigerant passing through a service truck. Sample forms are provided in Appendix III.

Virgin Refrigerant Use Log

Each time refrigerant is used, this form should be completed in addition to normal entries on the field service ticket. Documentation for usage of each refrigerant by the pound is scribed within the appropriate column. Documentation also includes all used refrigerant being returned for reclaim.

Contaminated or mixed refrigerant being returned is recorded. The cylinder or drum could contain all types of refrigerant. Weights stated are inclusive of all refrigerants, reflecting container Net Weight.

At the bottom of this form, an entry is made reflecting how many pounds of refrigerant was received over a given time period, according to type of refrigerant. It is suggested that the technician will not receive virgin refrigerant without first returning an empty drum for that type of refrigerant. Exception could be made, based upon a specific job requirement. Material requisition may be the primary means of transferring contained refrigerant. However, this entry will be a double check. All information is to be completed, including signature and date.

This form should be a multi-part device, with copies to respective areas of responsibility. It does not have a provided column for refrigerant lost accidentally or unintentionally. If that should happen, the form entitled "Accidental or Unintentional Venting Report" should be used.

Accidental or Unintentional Venting Report

It is not the intention of the Clean Air Act to penalize anyone if they accidentally vent refrigerant. The law states "good faith effort." However, for protection, institution of this form is recommended. It is further suggested that the customer sign any completed form that is applicable. The form may also be necessary to keep close accounting of all refrigerant used by the pound.

If an unintentional venting occurs, *complete all of the details* on this form. Each time there is an incident, a separate form is completed. In addition, notify your central office immediately.

Completion of the form should be considered part of the service call. There is no need to rush; the important thing is that it be completed fully and accurately.

The EPA will not tolerate careless practices or disregard for the law, which results in venting refrigerant.

Refrigerant Removal Incident and Leak Repair Report

This form should be completed whenever refrigerant has to be removed from a system. If a system is pumped-down for repairs, this does not have to be completed, provided the system low side has zero PSI pressure within it. In addition, no refrigerant can be vented to the atmosphere. In all pertinent cases, this form should be thoroughly completed.

Intent for repairing leaks is to eliminate them now and make sure they do not recur. The customer must be made aware of what is happening. Informative brochures, such as the ACCA's *Air Conditioning, Refrigerant, and the Ozone Layer: What Homeowners Need to Know*, can be furnished to your customer to promote awareness of regulations and the importance of leak elimination. Do not just repair the leak. Find out what caused it to occur and remedy the cause.

This can be a multi-part form, with copies to respective areas of responsibility. Complete all information as requested, for your protection. If an incident is reported by an outside party, this form is your proof and protection that proper procedures are being followed.

Notes

Introduction

Who Needs Small Appliance Certification?

Any technician who services sealed equipment with a refrigerant charge of five pounds or less is required to hold a Type I (Small Appliance) or Universal Certification by November 15, 1994. **Once certified, it will be the certified technician's responsibility to comply with any changes in the law.**

The EPA defines a small appliance as: Any of the following products that are fully manufactured, charged, and hermetically sealed in a factory with five pounds or less of refrigerant.

Small appliances include refrigerators and freezers designed for home use, room air conditioners (including window air conditioners and packaged terminal air conditioners), packaged terminal heat pumps, dehumidifiers, under-the-counter ice makers, vending machines, and drinking water coolers.

The EPA defines a Technician as: Any person who performs maintenance, service, or repair that could reasonably be expected to release class I (CFC) or class II (HCFC) substances into the atmosphere, including but not limited to installers, contractor employees, in-house service personnel, and in some cases, owners.

Refrigerant Recovery

Refrigerant RECOVERY is defined as: Removing refrigerant in any condition from a system in either an **active**, or **passive** manner, and storing it in an external container without necessarily testing or processing it in any way.

Before starting a recovery, a technician must always know what type of refrigerant is in the refrigeration system.

Active Recovery

"ACTIVE" or SELF-CONTAINED RECOVERY is: Recovering refrigerant from a system through the use of a self-contained recovery machine. A self-contained recovery machine has its own built-in compressor.

Active recovery devices **do not** use the refrigeration or air conditioning system's compressor to aid in recovery. Active recovery devices can be designed to recover liquid, vapor, or a combination of both. If the active recovery device is designed to remove vapor only, vapor should be removed from the high side of the sealed refrigeration system to avoid getting liquid.

If the refrigeration system happens to have service valves, avoid trapping liquid refrigerant between service valves while recovering.

In active recovery, the refrigerant is pumped into a Department of Transportation (DOT) approved recovery cylinder. In fact, portable, refillable cylinders that come with recovery equipment must meet the Department of Transportation standards.

Whenever refrigerant is transferred to a DOT approved pressurized cylinder in an active recovery, a safe filling level of the recovery cylinder must be obtained. **The recovery cylinder can only be safely filled to the 80% liquid level.** Below are three of the most common methods to accomplish this 80% safe liquid filling level.

❑ Mechanical float and electric switch (80% stop-fill switch). Refer to Section Nine under Stop-Fill Switch in this Manual.

❑ Electronic sensors (thermistors)

❑ Weighing refrigerant in the recovery tank with a scale before and after each transfer. This method is used with sensorless tanks.

Passive Recovery

"PASSIVE" or SYSTEM DEPENDENT RECOVERY is: Recovering refrigerant from a system employing the refrigeration or air conditioning system's internal pressure and/or system's compressor as an aid in the recovery process.

Passive recovery is probably going to be used most by technicians servicing small appliances. **However, when using passive recovery, the refrigerant must be recovered in a non-pressurized container (see Figure 7-1).** In fact, any device that is a non-pressurized storage vessel for capturing refrigerant can be considered a passive recovery device.

NOTE: *Recovery devices should be leak checked with a refrigerant leak detector on a "daily" basis.*

Speeding up any Recovery Process

Speeding up the recovery process can be accomplished in several ways. Below are some commonly used methods.

❑ Heat the system being recovered with a heat gun, heat lamp, or any source of heat that doesn't have an open flame. Heating the compressor's crankcase is a good place to start. Turning on a defrost heater to

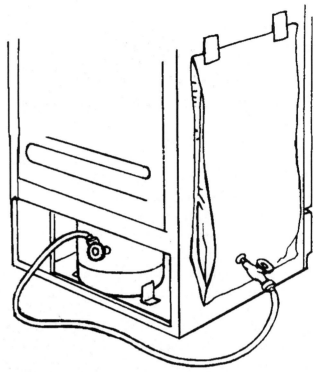

Figure 7-1.

release any trapped liquid refrigerant will surely help the recovery process.

❑ Gently strike compressor with a rubber mallet.

❑ Place the recovery cylinder in ice or cold water to lower its pressure.

❑ Recover refrigerant in the highest possible ambient to assure the highest system pressure.

Low Loss Fittings

Recovery equipment manufactured after November 15, 1993 must be equipped with low loss fittings. There are numerous types of low loss fittings on the market. Low loss fittings are devices which can be manually closed off or that automatically close off when they are disconnected from the recovery device and/or small appliance (Figure 7-2).

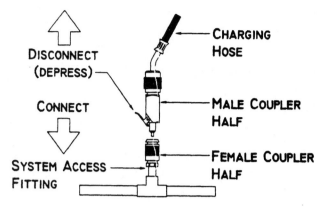

Figure 7-2.

Schrader Valves

Schrader valves on recovery equipment must be periodically checked and examined for cracks, breaks, and bends (Figure 7-3). If schrader valves are found to be defective or damaged, replace the schrader valve core to prevent system leakage. Also, all schrader valves must be capped to prevent leakage and accidental depression of the valve core.

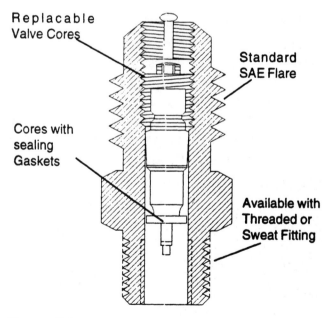

Figure 7-3.

Piercing Valves and Service Apertures

One of the main concerns of service technicians is where and how to access a small appliance to check pressures or recover refrigerant. The most common and widely accepted method to access a small appliance is with a piercing valve. Piercing valves come in clamp-on or solder-on types. Remember, as of July 1, 1992, it is illegal to simply cut the system tubing and let CFC or HCFC refrigerants escape to the atmosphere. In fact, as of July 1, 1992 it is illegal to vent CFC and HCFC refrigerants in any way.

As of November 15, 1995, it is illegal to vent alternative refrigerants including HFC-134a and refrigerant blends. Even venting regulated CFC or HCFC refrigerants off the top of a graduated charging cylinder is a venting violation. These refrigerants must be recovered when filling a graduated charging cylinder.

Piercing valves gain access to refrigeration systems by driving a sharp, needle-like steel instrument through the copper or aluminum tubing. When the needle is backed out of the tubing, the system can be accessed.

Clamp-on type or solderless piercing devices should be used only when temporary access is needed. They have a gasket material made of either rubber, Neoprene, or a certain composite that will deteriorate in time when exposed to hot environments and cause leaks. **Clamp-on type piercing valves should be removed once recovery is finished**. Always leak check these access valves before piercing the tubing and attempting a refrigerant recovery. It is strongly recommended that any type of piercing valves be used on copper or aluminum tubing only.

Solder-on type piercing valves are actually soldered on the system's tubing and then a steel piercing pin is driven into the tubing. Always leak check the solder joint for leaks before piercing the system's tubing by pressurizing the saddle valve body. Care must be taken when soldering or brazing the saddle valve body onto the tubing. The tubing cannot get so hot as to rupture causing a blowout of refrigerant and possible injury. Solder-on valves can be permanently left on the system after system access.

NOTE: *If after accessing a small appliance with a piercing valve and the system's pressure is 0 psig, do not attempt recovery. This a good indication that there is a leak in the system. The recovery device will only take in air, non-condensable, and water vapor. A 0 psig reading could also indicate that the line has not been successfully pierced yet.*

Some small appliances have a small service aperture or process stub or tube protruding out from the compressor (Figure 7-4). These are often referred to as process stubs. These apertures may protrude from the high side of the system but are mainly seen on the

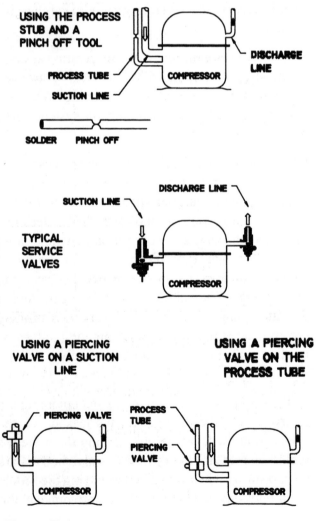

Figure 7-4.

compressor for low side access. They are usually a straight piece of tubing that can be pierced with a piercing valve for system access. **When the systems compressor is inoperative, both low and high side system access is recommended to speed up the recovery process and to achieve the required vacuum levels.**

When employing a **passive recovery process** when the system's compressor is inoperative, both low and high side access is required for refrigerant recovery.

When employing a **passive recovery process** when the system's compressor is operative, you should run the system's compressor and recover the system's refrigerant from the high side of the system.

Vacuum Pumps

Because of their internal construction, vacuum pumps cannot handle pumping against anything but atmospheric pressure.

Vacuum pumps should be used only to recover refrigerant in "passive" recovery devices where refrigerant is recovered into "non-pressurized" containers.

Vacuum pumps are very efficient when recovering refrigerant into a non-pressurized container. Their abilities to create very low vacuums will assist in recovering most of the refrigerant in a small appliance, especially when the system's compressor is inoperative. Vacuum pumps are very effective in separating refrigerant dissolved in the compressor's crankcase oil.

If a technician uses a vacuum pump that is too large for the system, the moisture in the system may freeze due to a quick drop in pressure. Dry nitrogen may then be introduced to the system to bring up the pressure and avoid freezing of the moisture. Also, if the system is under a deep vacuum, the compressor should never be energized. This will prevent motor winding damage from occuring due to a short circuit of the motor terminals inside the compressor.

After recovery and service of the equipment, evacuation using a vacuum pump is suggested for dehydration of a refrigeration or air conditioning system. Moisture left in the system will cause acids to form. During evacuation, the system's vacuum gauge should be located near the system's tubing and as far from the vacuum pump as possible. For any given system, the vacuum pump capacity in CFM and the vacuum pump's suction line diameter will determine how long it will take for the system to be dehydrated. Always measure the final vacuum with the vacuum pump isolated from the system and the vacuum pump turned off. If the system is to be recharged with refrigerant, never heat the refrigerant storage tank or the recovery tank with an open flame. Instead, partly submerge the tank in warm water to bring up the tank pressure. Heating the tank with an open flame may cause the refrigerant in the tank to decompose, forming harmful by-products.

When Not to Recover

❑ When access is gained and the system is found to be at atmospheric pressure, 0 psig. This is an indication of a leak and a recovery attempt will only contaminate the system and recovery device with air, non-condensables, and water vapor from the atmosphere.

❑ Any small appliance manufactured before 1950 if containing:
  ❍ Sulfur dioxide
  ❍ Methyl chloride
  ❍ Methyl formate

❑ Refrigerants found in absorption systems in campers and recreational vehicles such as:
  ❍ Water
  ❍ Ammonia
  ❍ Hydrogen

❑ When R-22 is mixed with nitrogen as a trace gas for electronic leak checking. In this case, the mixture of nitrogen and trace of R-22 can be vented to the atmosphere

because the ozone-depleting compound R-22 is not used as a refrigerant.

Note: *A technician may not avoid recovering refrigerant by adding nitrogen to a charged system. The system must be evacuated to the appropriate level in Table 7-1 before adding nitrogen.*

❑ When nitrogen is used as a holding charge. Nitrogen can be vented to the atmosphere.

❑ When a strong acidic or pungent odor is detected when servicing or accessing a small appliance. These are sure signs of a compressor burnout and the recovery equipment can be damaged and/or contaminated.

Table 7-1

SMALL APPLIANCE RECOVERY EFFICIENCY REQUIREMENTS		
RECOVERY EFFICIENCIES REQUIRED	RECOVERED PERCENTAGES	INCHES OF MERCURY VACUUM
For Active & Passive equipment manufactured after November 15, 1993 for service or disposal of small appliances with an operative compressor on the small appliance.	90%	4*
For Active & Passive equipment manufactured after November 15, 1993 for service or disposal with an inoperative compressor on the small appliance.	80%	4*
For grandfathered active and passive equipment manufactured before November 15, 1993 for service or disposal with or without an operating compressor on the small appliance.	80%	4*

*ARI 740-1993 Standards

Recovery Requirements & Levels

Recovery levels are set by law! In fact, any recovery equipment used for small appliances which has been manufactured after November 15, 1993 must be certified by an EPA-approved laboratory.

Within 20 days of opening a new business, you must register (or certify) to the EPA that you have recovery equipment that is able to recover at least 80% of the refrigerant charge, or achieve a 4 inch mercury vacuum under ARI Standard 740-1993. (See Appendix III.)

As of December 31, 1995, the Clean Air Act requires that the production of CFCs be stopped.

Leak Checking

The final EPA rule does not address the repair of leaks for small appliances. In fact, it is **not** mandatory to repair leaks on small appliances. However, an environmentally conscious service technician should make every effort to find and repair the leak.

Once a small appliance has had its refrigerant recovered and is suspected of leaks, it should be leak tested. **Dry nitrogen is the recommended method of leak checking R-134a systems. R-134a is a substitute for R-12 in small appliances, but it is not a direct drop-in refrigerant for R-12.**

Whenever dry nitrogen is used for pressurized leak checking, the use of a pressure relief valve and pressure regulator must be used on the nitrogen tank. Also, pressure relief valves must **never** be installed in series. Remember, nitrogen is an extremely high pressure gas and must be regulated.

Always check the design low side test pressure on the system's data nameplate before pressurizing a system with dry nitrogen. Never exceed this value.

Shipping of Refrigerant Cylinders

When refrigerant or recovery cylinders are being transported in any way, they must be transported in a vertical, upright position.

Refrigerants should never be mixed within the cylinder. When refrigerants are recovered from an appliance, they must be recovered in their own clearly marked cylinders. If mixed, the refrigerant must be taken to a reclaimer. The reclaimer may or may not accept it. When the reclaimer receives a cylinder of refrigerants that have been mixed, they will either refuse to reclaim the refrigerant mixture, or agree to destroy the refrigerant mixture and charge

a fee. It is the technician's responsibility to correctly handle the mixed refrigerants and not vent it to the atmosphere. Refrigerant that is returned to a reclaimer must have a label identifying the type of refrigerant in the cylinder.

The shipping cylinders must be properly labeled and tagged to meet local, state, and federal regulations. Recovery cylinders must be colored gray with a yellow shoulder.

In summary, all refrigerant cylinders shipped must:

❑ Have proper labeling

❑ Meet Department of Transportation (DOT) standards

❑ Have the proper shipping paperwork which specifies the number of cylinders of each gas.

Disposal of Small Appliances and Disposable Cylinders

Throwing away an old appliance with the refrigerant intact is illegal. If you are involved with disposing of an old appliance, you must:

❑ Recover the remaining refrigerant from the appliance.

❑ Verify by a signed statement from whom you obtained the appliance that the refrigerant has been recovered. The date of recovery is also needed.

When a technician is finished using a disposable cylinder, the residual vapor in the cylinder must be recovered.

The valves should then be removed and the tank marked empty. The cylinder should then be sent to a metal scrapper. The reuse of a disposable cylinder is illegal.

Safety

Because refrigerants are heavier than air, they can displace the air and cause suffocation. Self Contained Breathing Apparatus (SCBA) must be used if persons are exposed to large refrigerant leaks, spills, or dangerous concentrations. If no Self Contained Breathing Apparatus is available, the refrigerant leak or spill area must be ventilated.

Active recovery devices may generate high pressure conditions if there is excessive air in the recovery cylinder or the recovery cylinder's inlet valve is not opened. Safety goggles or glasses and butyl-lined gloves must be worn when operating recovery devices especially when connecting or disconnecting hoses. For more information on safety, refer to Section Nine of this manual.

Certain refrigerants are flammable and toxic. Refer to the Safety Group Classification System section in Section Nine of this manual.

When refrigerants are exposed to high temperatures such as hot flames or metals, HCFC and CFC refrigerants can decompose into hydrochloric and hydrofluoric acids and Phosgene gas. These gases are very harmful and toxic especially when inhaled.

When leak checking, never pressurize a system with oxygen or compressed air. When exposed to the oxygen in the air, the oil in the refrigeration system will oxidize rapidly and dangerous pressures can be generated to the point of explosion. Nitrogen is the only gas that can be safely used to pressure test a system when refrigerants and oils are present.

Notes

Notes

❑ Process Definitions
❑ Refrigerant Containment Options
❑ Refrigerant Recovery Guidelines
❑ Changing Recovery Cylinders During Recovery
❑ Refrigerant Recycling
❑ Equipment Used for Recovery and Recycling
❑ Recovery and Recycling Machine Standards

Process Definitions

Differences between the terms recover, recycle, and reclaim must be completely understood and properly used within the industry.

Recover

To remove refrigerant in any condition from a system in either an *active* or *passive* manner, and store it in an external container without necessarily testing or processing it in any way.

The recovery process has been used in the refrigeration field for many years. Typically when a system was repaired and the refrigerant was not contaminated, the refrigerant was reused in the repaired system.

Recycle

To reduce contaminants in used refrigerant by oil separation, non-condensable removal, and single or multiple passes through devices that reduce moisture, acidity, and particulate matter, such as replaceable core filter-driers.

This term usually applies to procedures implemented at the field job site or in a local service shop.

There are many refrigerant recycling units available for all aspects of the refrigeration industry. The service technician must be trained in operation of these units, and know how to apply them at the job site or service shop.

Education and proper test equipment will be the key to successful recycling programs.

Reclaim

To process used refrigerant to new product specifications by means which may include distillation. Chemical analysis of the refrigerant is required to assure that appropriate product specifications are met.

This term usually implies the use of processes or procedures available at reprocessing or manufacturing facilities or field equipment that is capable of achieving the ARI-700 specifications for fluorocarbon refrigerants.

The EPA requires use of either recycling or recovery equipment. In most instances system refrigerant is still good for reuse. Recovery and replacement can be accomplished with minimal time and effort. It is important that the service technician take precautions when removing and transferring the refrigerant to avoid contamination.

When equipment operation indicates contamination or refrigerant deficiency, refrigerant can be recovered and recycled on site. If contamination is severe or exacting standards must be met, refrigerant must be reclaimed. Once refrigerants are contaminated or mixed, complicated procedures must be used for separation. Reclamation by refrigerant distillation can separate some refrigerants. If refrigerant is badly contaminated and cannot be separated by distillation, it must be sent to an authorized treatment facility for destruction. The method of destroying mixed refrigerants is to incinerate the refrigerant in a manner that will allow the fluorine to be captured. This process is very expensive.

The agency does allow refrigerant to be returned to air conditioning and refrigeration equipment with or without treatment, and allows transfer of recycled refrigerant between equipment owned by the same entity. But the contamination levels of those refrigerants must not exceed the maximum contaminant levels of recycled refrigerants as shown in the table below. **Refrigerant changing ownership must be fully reclaimed to ARI Standard 700 specifications.**

Maximum Contaminant Levels of Recycled Refrigerants in Same Owner's Equipment			
Contaminants	Low-pressure Systems	R-12 Systems	All Other Systems
Acid content (by wt.)	1.0 ppm	1.0 ppm	1.0 ppm
Moisture (by wt.)	20 ppm	10 ppm	20 ppm
Non-condens-able gas (by vol.)	N/A	2.0 %	2.0 %
High-boiling res-idues oil (by vol.)	1.0 %	0.02 %	0.02 %
Chlorides by silver nitrate test	no turbidity	no turbidity	no turbidity
Particulates	visually clean	visually clean	visually clean
Other refrigerants	2.0 %	2.0 %	2.0 %

Note: To ensure that the recycling equipment maintains its demonstrated capability to achieve the above levels, it must be operated and maintained per the equipment manufacturer's recommendations.

Refrigerant Containment Options

Option 1 Recover and reuse without processing. (Put refrigerant back into the system without recycling it.)

Option 2 Recover and recycle. (Recycle the refrigerant and put it back into the system from which it was removed or back into a system with the same owner.)

Option 3 Recover and reclaim. Provide documentation that the reclaimed refrigerant meets ARI-700 Standard of purity and is chemically analyzed to verify that it meets this standard.

Option 4 Recover and destroyed.

Guidelines have been established to assist the servicing contractor in determining which of the four options should be chosen.
- ❏ Reason the system is being serviced
- ❏ Condition of refrigerant and system
- ❏ Equipment manufacturer policy
- ❏ Refrigerant cleaning capability of recycling equipment
- ❏ Feasibility and owner's preference

See the flowchart entitled "Used Refrigerant Guidelines" on page 98.

Refrigerant Recovery

Removal of refrigerant from a system can be accomplished by two basic methods, passive or active recovery. To comply with government regulations and best serve customer needs, time must be taken to evaluate the system and determine which method to employ.

Questions to be considered by the technician
- ❏ Is system compressor operable?
- ❏ Is this system accessible enough?
- ❏ Where is the liquid refrigerant within the system?
- ❏ What is the outside temperature?
- ❏ Will outside conditions have any effect?

If the system is not analyzed, recovery could take longer than necessary.

Passive Method
System-Dependent recovery or "Passive" recovery is recovering refrigerant from a system employing the refrigeration system's internal pressure and/or system's compressor as an aid in the recovery process. **System dependent equipment shall not be used with appliances normally containing more than 15 pounds of refrigerant.** To make it easier for technicians to recover refrigerant, the EPA is requiring the manufacturer to install a service aperture

or process stub for appliances containing Class I and II refrigerants. If a service technician uses Passive or System-Dependent recovery on a system with an inoperative compressor, the refrigerant must be recovered from both the low and high side of the appliance to speed the recovery process and to achieve the required recovery efficiency requirements. A vacuum pump can be used in this procedure. However, never discharge a vacuum pump into a pressurized container. Vacuum pumps cannot handle pumping against anything but atmospheric pressure. If the compressor is operative, refrigerant can be recovered from the high side only. In all passive recovery, the refrigerant must be recovered in a non-pressurized container. Whether the compressor is operative or not, gently striking the compressor with a wood or rubber mallet during recovery will agitate and release the refrigerant dissolved in the compressor's crankcase oil. This will aid in the recovery process. Contingent upon the following, refrigerant can be moved without damage to the compressor:

❑ An adequately sized receiver or condenser
❑ Adequate recovery containers
❑ Weight recording method
❑ Proper on-off controls
❑ Not exceeding container's maximum net weight

Active Method

The most common method of system refrigerant removal is through use of a certified self-contained recovery unit, often referred to as a recovery/recycling machine. These machines are capable of both liquid and vapor removal. The first goal should be to remove system refrigerant in liquid form for increased recovery efficiency. **A higher ambient temperature also facilitates more rapid recovery due to increased system internal vapor pressure**. Methods of recovery are as follows:

Example One: Reverse Recovery *(see Figure 8-1)*
The objective is to reverse recovery, taking vapor from the system, injecting remaining vapor into liquid side of recovery container. Attach hose to condenser outlet or receiver king valve. Liquid is to flow from the receiver king valve to the recovery container (see Liquid Recovery). Attach a hose from vapor valve of the recovery container to recovery unit inlet. Attach another hose from outlet of the recovery unit through gauge manifold to system vapor side. When all liquid is removed, reverse hoses on recovery unit (see Vapor Recovery) and draw remaining vapor from the system. Vapor valve on recovery container remains closed.

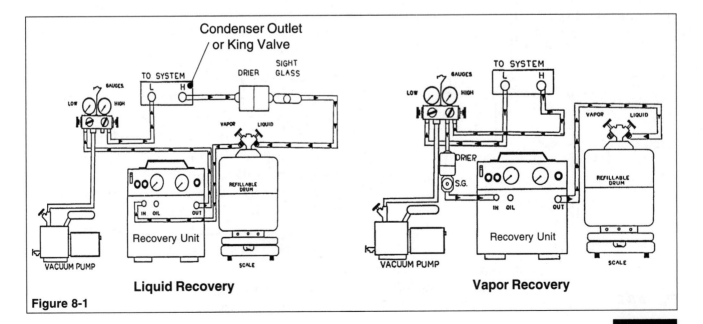

Liquid Recovery

Vapor Recovery

Figure 8-1

Example Two: Typical Recovery Procedure
(see Figure 8-2)

1. Connect the quick seal end of the hose to the **Liquid** valve fitting on the refillable tank. Open the **Liquid** valve.

2. Connect a manifold gauge set to the A/C system.

3. Connect the center hose from the manifold to the **Inlet** fitting on the recovery unit.

4. Open both the high and low side valves on the manifold.

5. Plug the recovery unit into a power source. The fan will start.

6. Turn **Valve 1** and **Valve 2** to the **Equalize** position.

7. After a few seconds, turn each of the valves to the **Recover** position.

8. Turn the power switch of the recovery unit on. The compressor will start.

9. When the **System Pressure Gauge** reaches the appropriate level of vacuum, close the high and low side valves of the manifold. Turn the power switch of the unit to the off position.

10. Recovery is complete.

If, during a recovery procedure, the refillable tank reaches the 80% capacity level for selected refrigerant type, **STOP** the recovery procedure. The refillable tank must be replaced at this time.

1. Turn off the Compressor Start switch. Fan will continue to run.

2. Close the tank valves.

3. Close **Valve 1** and **Valve 2** on the recovery unit.

CAUTION! Hoses contain pressurized refrigerants.

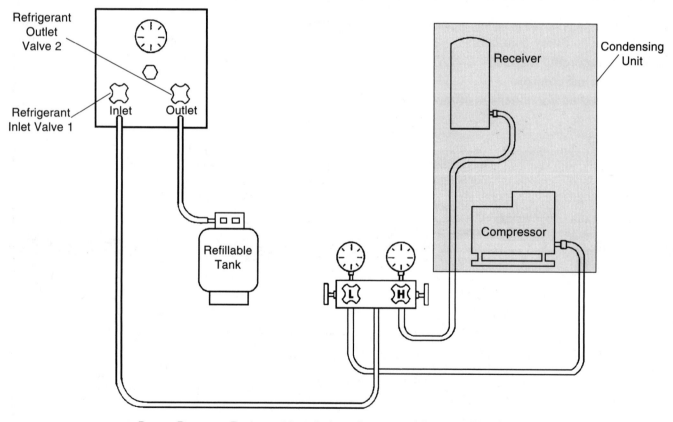

Figure 8-2.   Proper Recovery Equipment Installation - Courtesy of Robinair SPX Corp
(Other Typical Plumbing Diagrams — See Figure 8-3 and Figure 8-4)

Refrigerant Transition and Recovery Certification Program Manual

Typical Plumbing Diagrams

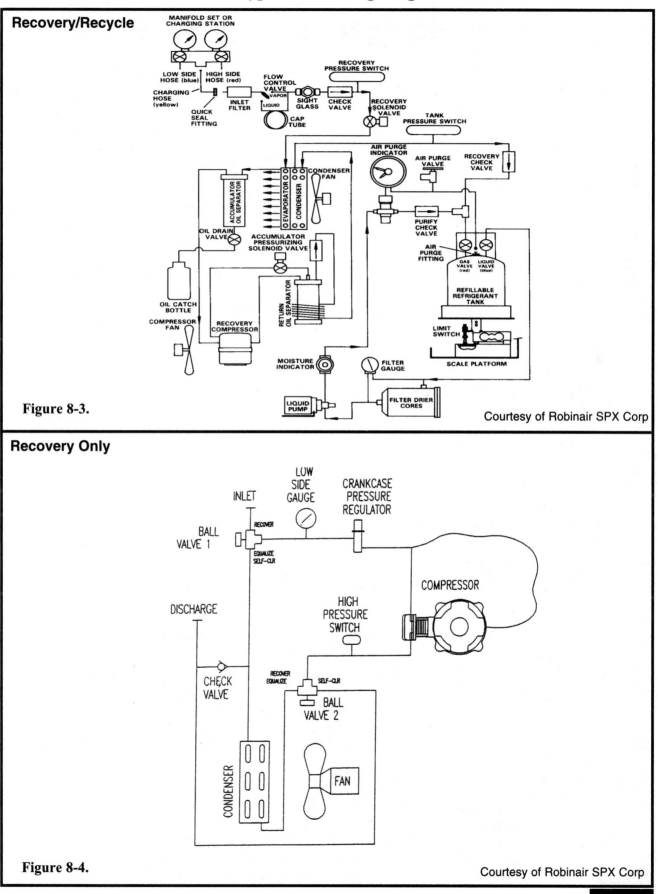

Recovery/Recycle

Figure 8-3.

Courtesy of Robinair SPX Corp

Recovery Only

Figure 8-4.

Courtesy of Robinair SPX Corp

Used Refrigerant Guidelines

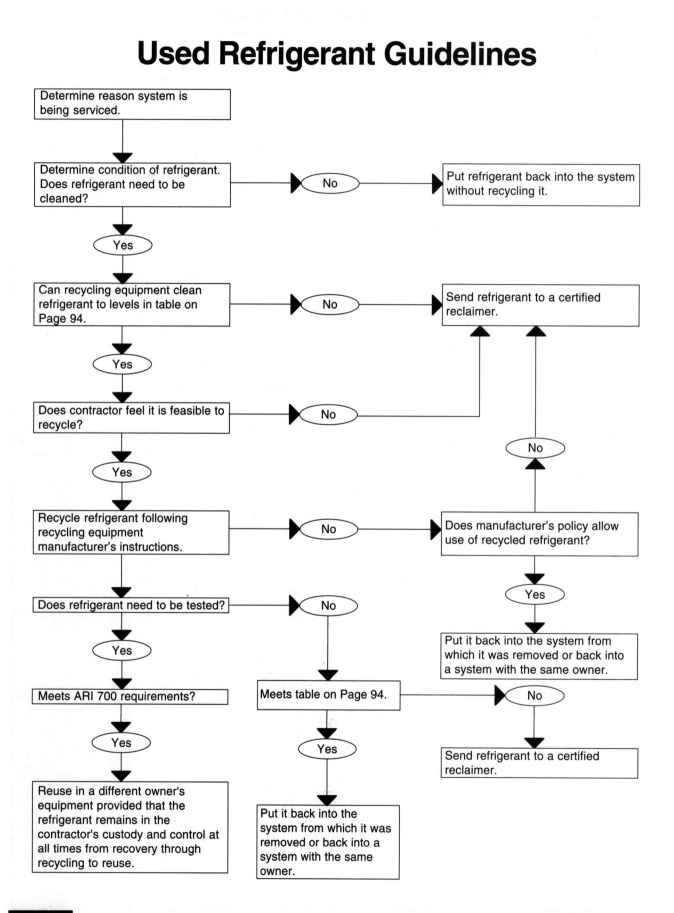

Determine reason system is being serviced.

Determine condition of refrigerant. Does refrigerant need to be cleaned? — No → Put refrigerant back into the system without recycling it.

Yes

Can recycling equipment clean refrigerant to levels in table on Page 94. — No → Send refrigerant to a certified reclaimer.

Yes

Does contractor feel it is feasible to recycle? — No

Yes

No

Recycle refrigerant following recycling equipment manufacturer's instructions. — No → Does manufacturer's policy allow use of recycled refrigerant?

Yes

Does refrigerant need to be tested? — No

Yes

Put it back into the system from which it was removed or back into a system with the same owner.

Meets ARI 700 requirements?

Meets table on Page 94. — No

Yes

Yes

Send refrigerant to a certified reclaimer.

Reuse in a different owner's equipment provided that the refrigerant remains in the contractor's custody and control at all times from recovery through recycling to reuse.

Put it back into the system from which it was removed or back into a system with the same owner.

4. Disconnect the hose(s) from the refillable tank. Replace refrigerant tank with an empty tank under vacuum of the same refrigerant type. Do not cross-contaminate refrigerants.

5. Reconnect hose(s) to empty refrigerant tank.

6. Open tank valves.

7. **Equalize** and restart recovery unit.

Refrigerant Recovery Guidelines

❑ Refrigerant can be removed from a system as a liquid for increased recovery efficiency.

❑ Refrigerant, when removed in the vapor stage, will minimize the loss of oil from the refrigerant system.

❑ If the refrigerant is not moving through the recovery unit, the liquid line valve is not open, the valve on the recovery unit is not positioned correctly, the hose connections are leaking, or there is a faulty connection at the manifold gauges.

❑ Refrigerant recovered in the recovery cylinder must be checked for non-condensables after every recovery. Compare pressure reading in the recovery tank to pressure/temperature chart. **A higher pressure reading in the recovery tank than the pressure/temperature chart indicator for ambient is evidence of non-condensables in the recovery cylinder.**

❑ If, after accessing a small appliance with a piercing valve and the system pressure is 0 psig, do not attempt recovery.

❑ Never try to recover sulfur dioxide, methyl chloride, methyl formate, ammonia, or hydrogen.

Note: It is considered dangerous to use CFC or HCFC recovery equipment to recover ammonia, hydrocarbons, or chlorine. However, users of hydrocarbon, ammonia, and pure chlorine refrigerants must continue to comply with all other applicable federal, state, and local restrictions on emissions of these substances.

❑ Lowering the temperature of the recovery cylinder using ice or ice water will cause a pressure difference between the system and the recovery cylinder, speeding up the recovery time.

❑ When recovering refrigerant from a system where the condenser is below the receiver, refrigerant should be removed from the condenser outlet, receiver outlet, metering device to the evaporator inlet, or the lowest point of the liquid line that has a service valve.

❑ Removal rates for recovery units will vary due to refrigerant types, ambient temperatures, and system connections.

❑ If recovery is slow, switch to an empty tank under a vacuum, a cooled tank, or a tank with less refrigerant than the existing tank.

❑ Refrigerant hoses should be of top quality (low permeation) with good sealing gaskets and automatic shut-off valves.

❑ Do not use a recovery unit on a system that is known to have water mixed with the refrigerant (water cooled condenser with a tube leak).

❑ Recovery units usually have a sight glass that will indicate the condition of the refrigerant after passing through the oil trap and both filters.

❑ Moisture indicators are usually found within the sight glass to indicate moisture content of the refrigerant or condition of the filters.

❑ Do not use a recovery machine to recover refrigerant from a system known to have a burned out compressor. The recovery equipment could be damaged. Use a method such as a chilled refrigerant recovery tank or a tank in a deep vacuum. Remove as much refrigerant as possible. The remaining will be a DeMinimus release.

Changing Refrigerant Types and Filters

Refrigerant must be removed from the recovery unit before changing refrigerant or filters through use of a second recovery system. De minimis release is permitted.

Many new recovery units are now designed to be self-clearing. The self-clearing (purge) feature allows the recovery unit to be used with almost all common refrigerants as long as it is designed to handle the operating pressures of the refrigerant.

Any filters installed as a part of the recovery unit or any inline filters should be replaced after 200 pounds of processing, or as required by the manufacturer of the recovery unit. (See Figure 8-5 and Figure 8-6.)

Most inline filters are directional and must be oriented to refrigerant flow. Refrigerant flow is indicated by an arrow on the filter body.

Filters should be replaced sooner if any of the following conditions exist:
❑ High acid level in compressor oil.
❑ Moisture indicator shows wet or caution when recovery unit is not in operation.
❑ Flow rate is significantly reduced.

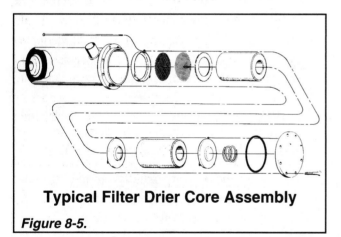

Typical Filter Drier Core Assembly

Figure 8-5.

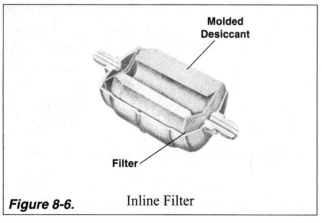

Molded Desiccant

Filter

Figure 8-6.   Inline Filter

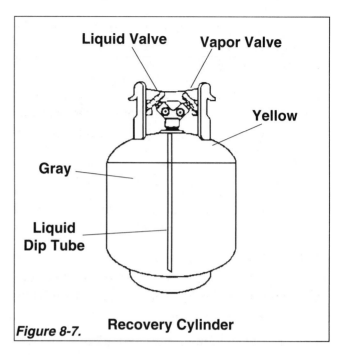

Liquid Valve   Vapor Valve

Yellow

Gray

Liquid Dip Tube

Figure 8-7.   **Recovery Cylinder**

Changing Recovery Cylinders During Recovery

A recovery cylinder (Figure 8-7) must have some way of ending the recovery process when it reaches 80% of the cylinder's capacity. **Float device in the cylinder, electronic thermistors built into the cylinder, and weight are all methods to terminate recovery when this level has been achieved.** See Section Nine, page 116.

The cylinder must be replaced before the unit can be started. The replacement cylinder can contain refrigerant but must not be full. **Make sure not to mix refrigerants.**

Steps to avoid mixing refrigerants:
❑ Properly clean the recovery unit.

❑ Dedicate oil type recovery units to a specific refrigerant.

❑ Use a self-clear recovery unit.

❑ Dedicate a cylinder to a specific refrigerant.

❑ Suspect refrigerant should be tested (use pressure/ temperature chart) before consolidating into larger containers and before attempting to recycle or reuse.

- ❑ Use refrigerant identifiers and air detection/purge devices that are now available for suspect refrigerant testing.

- ❑ Assure that containers are free of oil and other contaminants. Liquid recovery may increase the likelihood of contaminated cylinders because of oil entrainment.

- ❑ Keep appropriate records of refrigerant inventory.

- ❑ Cylinders used for recovery and/or recycled refrigerant should be marked appropriately.

Refrigerant Recycling

Recycling of refrigerant removes contaminants such as moisture, acid, non-condensable gases, particulate, and high-boiling residues.

Most system contamination is entrapped in the compressor crankcase oil charge. Although much remains within system components, particulate, sludge, and organic acids are removed when oil is separated from recovered refrigerant. That which is left is residual within the system. If the compressor is removed, most system contamination is also removed. As refrigerant is recycled, oil is separated and refrigerant is strained through drier cores prior to insertion into the recovery cylinder.

The primary function of the refrigerant recycling filter/drier is to remove moisture, while the secondary function is to remove acid, particulate matter, sludge, and suspended varnish. If oil samples prove satisfactory, refrigerant stored in a recovery cylinder should be filtered once more before insertion into the system through an in-line filter/drier installed in the charging line.

If an oil sample proves acid contamination, it is in the best interest of system integrity that the refrigerant be recovered by the passive method and then reclaimed to ARI-700 Standards. Depending on severity, if contaminated refrigerant is not reclaimed and instead recharged into the system, follow procedures for a system with a severely burned out compressor. Install an over-sized activated alumina liquid line drier and a properly sized suction line burn-out drier.

After the unit is put into operation, monitor pressure drop across the suction line drier and replace as necessary. It is also advisable to replace the liquid line drier when the suction line drier requires replacement. When acid tests of oil samples prove clean, final replacement of the suction line drier with a suction line filter is recommended. Liquid line drier may also be replaced to assure continued protection.

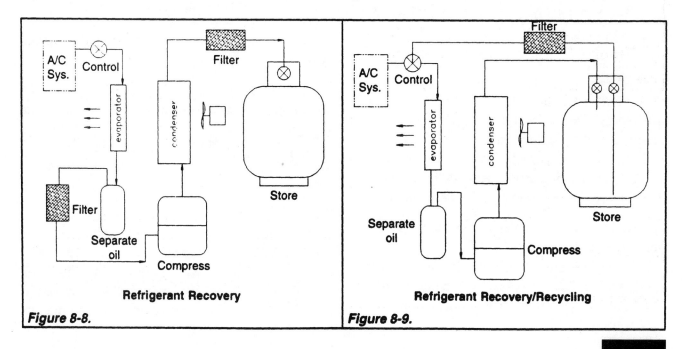

Figure 8-8.   Refrigerant Recovery

Figure 8-9.   Refrigerant Recovery/Recycling

Table 1		
Required Levels of Evacuation of Air conditioning, Refrigeration, and Recovery/Recycling Equipment (except for small appliances, MVAC's and MVAC like equipment) Inches of Hg Vacuum (Relative to Standard Atmospheric Pressure of 29.9 Inches Hg.		
Type of Appliance	Using recovery or recycling equipment manufactured or imported before November 15, 1993	Using recovery or recycling equipment manufactured or imported on or after November 15, 1993
HCFC-22 appliance, or isolated component of such appliance, normally containing less than 200 pounds of refrigerant.	0	0
HCFC-22 appliance, or isolated component of such appliance, normally containing 200 pounds or more of refrigerant.	4	10
Other high pressure appliance, or isolated component of such appliance, normally containing less than 200 pounds of refrigerant. (CFC-12, -500, -502, -114)	4	10
Other high pressure appliance, or isolated component of such appliance, normally containing 200 pounds or more of refrigerant. (CFC-12, -500, -502, -114)	4	15
Very high pressure appliance. (CFC-13, -503)	0	0
Low pressure appliance. (CFC-11, HCFC-123)	25	25 mm Hg. Absolute**
Note: MVAC = Motor Vehicle Air Conditioning Small Appliances = Manufactured, charged and hermetically sealed with 5 lb or less of refrigerant.		****25mm Hg absolute = 25,000 microns = 29 in Hg √ gauge (vacuum)**

If the entire refrigerant charge must be recycled, an oil separator is recommended. Many recovery/recycling equipment manufacturers provide two oil separators within their available equipment. Oil is isolated from the refrigeration system by vapor and liquid separation (velocity and direction change), or by distillation. Nearly all of the oil is removed from the refrigerant being processed with this apparatus, leaving only the compressor oil charge behind.

Equipment Used for Recovery and Recycling

A recovery machine in simplest form consists of a control valve (vapor passage or liquid metering device), oil separator, evaporator, compressor, condenser, and a container for holding the recovered refrigerant as shown in Figure 8-8.

Options found in Figure 8-9 show how the system can be operated as a recover/recycle machine with continuous operation after the recovery has been accomplished. Add a sight glass with a moisture indicator to the system and a filter designed for recycling and the refrigerant condition can be monitored.

Termination of the recovery process is generally accomplished with automatic low pressure controls. After initial shut down, residual evaporation of refrigerant in the system may develop a positive pressure. A second recovery can be performed to recover more refrigerant.

Recovery and recovery/recycling equipment is available in varying sizes, types, and prices. A great number of these units employ both filtration and distillation to process the refrigerant, while some are dedicated to filtration only. Standard models usually process 12, 22, R-502, and 134a refrigerants. There also are models that process 11, 113, 114, 123, and other refrigerants. Some models are capable of processing 12, 22, 502, 134a, 410A and Blends. Many recovery machines utilizing compressors containing oil can also be converted to accommodate blends through oil and refrigerant change. Some units are self-clearing. (Consult manufacturer for proper application.)

Processing time is usually calculated in lb./min. and range from 0.5 to 5 pounds of vapor, with the average unit processing approximately 1.5 to 3 pounds vapor/minute. Some manufacturers use liquid figures and indicate 8 to 12 pounds of liquid/minute. Be careful when determining capacities and make certain each manufacturer is using the same criteria when indicating capacity figures.

To cover all offerings of recovery equipment, there are units capable of 30 pounds, 300 pounds, and even 3600 pounds of liquid/minute. These are usually vacuum pumps, weighing from 500 to 2200 pounds and have an internal storage capacity of up to 10,000 pounds of refrigerant . Units of this capacity are used mostly for the recovery of 11, 113, 114, and 123 refrigerants.

Typical compressor types are hermetic, reciprocating, and rotary. Larger units are true vacuum pumps. Voltages are typically 115, 230, and 460.

Equipment weight is a factor for the service technician. The recovery/recycling units range from 29 pounds to 100 pounds. Some are in the range of 125 pounds to 160 pounds, while the larger units start at the 200 pound range. There are a variety of units each with a specific purpose. New type recovery units are now being offered by manufactures that include oil-less recovery units and oil-less recovery unit and vacuum pump combinations that can be used with most refrigerants and refrigerant blends.

Take precautions when selecting a recovery and recycling unit. Make sure it will perform the functions needed in your particular type of business. Always maintain your equipment as specified by the manufacturer. **Changing filters will be one of the most important functions, and is an excellent defense against contamination.**

Recovery Unit General Notes

❏ **Rising Pressure** - If system pressure rises above zero PSI on the charging manifold gauges, there is still refrigerant in the system. Smaller systems present a rapid rise in pressure, while larger systems take considerably longer.

Refrigerant still remains in the system if frosting occurs on any component. Recovery can resume when temperature rises.

❏ **Oil-less Refrigerant Recovery Units** - Almost all refrigerant recovery units now being made are of the oil-less design. This allows them to be used with all types of oils. Oil change and oil contamination is no longer a problem.

❏ **Draining the Oil Trap (if so equipped)** - Oil trap should be drained after every use and more frequently if the refrigerant quantity is large. Always measure oil being removed and replace that amount with new refrigeration oil.

❏ **Crankcase Sight Glass and Oil Outlet (if so equipped)** - Sight glass should be approximately one third to one half full. Do not use sight glass reading while unit is under a vacuum, as the sight glass might show a high oil level. If oil level drops below the sight glass window, add oil until the glass indicates one third to one half full. To drain oil, depress the valve core at the crankcase oil outlet. Pressure in the recovery unit will move oil out of the unit.

Follow manufacturer's recommendation for refrigeration oil.

- **Replacement Oil (if so equipped)** - Oil can be pumped directly into the crankcase against pressure by using a refrigeration oil pump. Oil can be added while the unit is operating. You may also pull new oil into the compressor by creating a vacuum and allowing the oil to flow into the crankcase.

- **High Pressure Safety Switch** - Most units have a high pressure safety switch which will shut down the unit at a predetermined pressure. This safety device will not prevent overfilling of the storage tank. In some cases, high ambient temperature will cause the equipment to shut down and the tank will need to be cooled to continue recovery.

Component Definitions

- **Low pressure gauge** - Indicates pressure in suction line of the recovery compressor.

- **High pressure gauge** - Indicates pressure in liquid outlet line.

- **Moisture indicator** - Shows condition of the refrigerant before entering the storage tank.

- **Inlet valve and fitting** - Controls flow into recovery unit. (Equipped with a check valve to prevent refrigerant flow out of inlet valve.)

- **Outlet valve and fitting** - Controls flow out of recovery unit. (Equipped with a check valve to prevent refrigerant flow into the outlet valve.)

- **Equalization valve** - Equalizes high and low side pressures within the recovery unit.

- **Vacuum valve and fitting** - Connection for vacuum pump and allows evacuation of hoses and unit.

- **Compressor crankcase sight glass** - Allows oil level to be monitored.

- **Tank cut-off socket** - Electrical connection on a storage tank containing a float switch.

- **Crankcase oil outlet fitting** - Allows the crankcase oil to be changed. (Keep capped when not being used.)

- **Oil trap outlet fitting** - Allows oil trap to be drained.

Recovery and Recycling Equipment Safety

- Units should only be operated by qualified refrigeration and air conditioning service technicians.

- Use only approved DOT 4BA (15-50 lbs.) and 4BW (100 lbs. and up) refillable storage cylinders. **Do not use disposable cylinders.**

- **Do not overfill refrigerant recovery cylinders (80% net maximum).**

- **Do not mix refrigerants.**

- Use the proper unit for the refrigerant being recovered.

- **Wear safety glasses and gloves.**

- Use in a well ventilated area (at least four air changes per hour).

Recovery and Recycling Machine Standards

- **Recovery Efficiency and Vacuum Levels Except for Small Appliance**
  Maximum recovery efficiency of recovery and recycling equipment is the percentage of refrigerant that the equipment is capable of recovering from an appliance, and is directly related to the depth of vacuum that the equipment can achieve. Required levels of evacuation are shown in the Table 1 on page 102.

- **Low-Loss Fittings**
  Low-loss fittings are intended to prevent not only the release of refrigerant, but an influx of air into the system. Devices to properly "deinventory," or empty the hoses of refrigerant

before disconnection are also required. Bleeding high side hoses into low side will minimize refrigerant losses.

❑ **Purge Loss from Recycling Equipment**
The amount of purge loss from recycling equipment is set at 3 percent. This value is based upon the total quantity of refrigerant being recycled.

❑ **Volume-Sensitive Shutoff**
In consideration of technical problems involved in establishing a requirement for a volume-sensitive shutoff switch, the agency has decided not to include this requirement in the rule. Instead, the EPA will include knowledge of proper cylinder filling in its requirements for technician certification.

Table 2

Small Appliance Recovery Efficiency Requirements		
Recovery Efficiencies Required	Recovered Percentages	Inches of Mercury Vacuum
For active and passive equipment manufactured after November 15, 1993, for service or disposal of small appliances with a operative compressor on the small appliance.	90%	4*
For active and passive equipment manufactured after November 15, 1993, for service or disposal with an inoperative compressor on the small appliance.	80%	4*
For grandfathered active and passive equipment manufactured before November 15, 1993, for service or disposal with or without an operating compressor on the small appliance.	80%	4*
*ARI 740 Standards.		

❑ **Recovery Machines Intended for use with Small Appliances**
Technicians servicing small appliances can employ either passive or active equipment. The required efficiency levels are shown in Table 2.

❑ **Recovery and Recycling Applied to Air Conditioning and Refrigeration Equipment (except for small appliances)**
The agency is requiring testing of recovery and recycling equipment by a third-party (i.e., ARI or UL). This is determined to be the most reliable method of obtaining an accurate and objective evaluation of equipment performance. While equipment certification alone does not guarantee minimized emissions, it does prevent leaking or ineffective equipment from entering the market. Home-built equipment was eligible for grandfathering prior to November 15, 1993 and it may continue to be used until no longer effective. After that date, home-built equipment is not allowed unless it is certified in the same way as equipment manufactured for sale.

❑ **Revocation of Certification**
In the event that equipment certification is revoked, the affected model of recycling or recovery equipment can no longer be manufactured. Similarly, in the event of revocation of a manufacturer's equipment registration, the equipment can no longer be used for recovery and/or recycling.

Grandfathering Provisions of Recovery and Recycling Machines
Equipment manufactured before November 15, 1993 may be grandfathered, providing it meets the standards of evacuation listed in Table 1 and Table 2.

Contractor Equipment Registration
Recovery and recycling equipment. Date of compliance was August 12, 1993. Contractors and others servicing refrigeration and air conditioning equipment, and those who dispose of appliances must certify that they have acquired or leased recovery equipment that meet EPA standards. This form can

be obtained through the EPA Hotline (800) 296-1996. A copy of this form can also be found in Appendix III of this manual.

Refrigerant Reclaimer Certification

Reclaimers are required to return refrigerant to the purity level specified in ARI 700, releasing no more than 1.5 percent of the refrigerant in the process. Purity must be assured using the laboratory protocol set forth in the same standard. Reclaimers must certify to the 608 Recycling Program Manager at EPA Headquarters within 90 days.

Reclaimer certification information can be obtained from:

608 Recycling Program Manager
Stratospheric Ozone Protection
Branch (ANR-445) EPA
401 M Street, SW
Washington, DC 20406

Certificates are not transferrable to a new owner.

Notes

Disposable and Returnable Refrigerant Containers

Refrigerants are packaged in both disposable and returnable shipping containers, commonly called "cylinders." They are considered to be pressure vessels, and therefore comply with federal and state law regulating transportation and usage of such containers.

Refrigerant manufacturers have voluntarily established common cylinder identifying colors:

R-11	Orange	R-13	Light Blue
R-12	White	R-123	Light Blue/Gray
R-134a	Light Blue	R404A	Orange
R-407C	Brown	R-410A	Rose
R-503	Aquamarine	R-22	Light Green
R-114	Dark Blue	R-502	Light Purple
R-113	Purple	R-500	Yellow
R-717, NH$_3$	Silver		

The shade of color may vary from one manufacturer to another, so you must verify contents by means other than color. Every refrigerant cylinder is silk-screened with product, safety, and warning information. Manufacturer technical bulletins and material safety data sheets are available upon request.

Even though cylinders are designed and manufactured to withstand saturated pressure of R-502 (the base refrigerant), **it is not recommended that any cylinder be repainted with a different color and used with another refrigerant.**

Saturated vapor pressure varies between refrigerants at given ambient temperatures. **Liquid refrigerant must be present within the closed container in order to read a pressure/temperature relationship, indicating saturated pressure.** As cylinder temperature increases, saturated pressure within the cylinder increases, corresponding with refrigerant temperature.

Pressure relief safety devices with releasing pressures preset to highest vapor pressure anticipated with R-502 are installed on every cylinder manufactured. They are of the frangible (rupture) disc style, or spring loaded relief integrated into the valve stem. Neither of the types are adjustable or are to be tampered with.

Disposable Cylinders

Department of Transportation (DOT) specifications require that disposable refrigerant cylinders be rated for a service pressure of 260 psi, and leak tested at 325 psi. Under laboratory test, one cylinder per

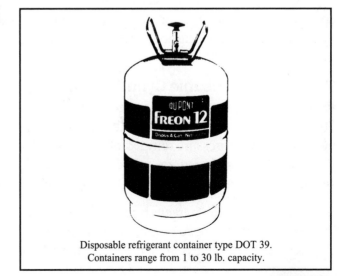

Disposable refrigerant container type DOT 39.
Containers range from 1 to 30 lb. capacity.

thousand produced is pressurized to point of failure. The cylinder must not rupture below 650 psi.

Disposable cylinders can be called "one trip" cylinders. Constructed of common steel, they do oxidize. Rust weakens the cylinder to where the wall and seams can no longer tolerate pressure and contain gasses. They also have a single acting valve, constructed of plastic and located at the top. Handles are provided, which can serve as rests for inverted liquid access from the cylinder.

Disposable cylinders are to be stored in dry locations to prevent corrosion, and transported carefully to prevent abrasion of painted surfaces. They are not to be refilled. The penalty for transporting a refilled disposable cylinder is a fine up to $25,000 and five years imprisonment.

Disposal of disposable cylinders is accomplished by recycling as scrap metal. When the cylinder is empty, assure that all pressure is released to zero psi. The cylinder should be made useless for any purpose.

Never allow used cylinders with residual refrigerant to sit at the job site. If the smallest amount of liquid is present, saturated vapor pressure will still exist. An abandoned cylinder could explode under direct heat from any source, especially if walls are weakened by corrosion.

The DOT has regulatory authority over all hazardous materials in commercial transportation. Disposable containers manufactured for CFCs are designed to Specification 39; hence, disposables are often referred to as "DOT 39s."

Returnable/Reusable Cylinders
Returnable cylinders meet DOT specification 4BA-300, with a water capacity of 122.7 lbs. Low boiling point, high vapor pressure refrigerants such as R-13 and R-503 are supplied in cylinders with DOT specification 3AA-1800, or 3AA-2015 respectively.

These cylinders are characterized by a combined liquid/vapor valve, located at the top. A dip tube feeding the liquid valve is immersed to the bottom to facilitate liquid removal without inverting the tank. Refrigerant can be removed in gas or liquid phase, through selection of either valve.

The large returnable cylinders possess a stamp on the shoulder which generally furnishes the following information:

- **OWNERS NAME (abbreviated)**
- **DOT SPECIFICATION NUMBER (for the cylinder)**
- **SERIAL NUMBER (of the cylinder)**
- **TEST DATE (month and year)**
- **MANUFACTURER'S SYMBOL**
- **WATER CAPACITY (in pounds)**

Cylinders Used in Recovery
According to DOT/ARI guidelines, all cylinders for used recovered refrigerant shall ultimately be **painted gray with the top shoulder portion painted yellow**. Proper identification is found on labeling. In all cases, the protective cap which screws securely to the cylinder body is painted gold. For recovery purposes, use only cylinders which are identified for used refrigerant. Never use a cylinder which came from the factory holding the original refrigerant.

Many recovery cylinders have 80 percent stop-fill devices incorporated in the tank. They have reed switches which open on the rise of liquid. When liquid reaches an 80 percent level in the recovery tank, a circuit is opened and the recovery machine is turned off.

Safe Handling of Recovered Refrigerant
❑ Become very familiar with your recovery equipment. Read the OEM manual and apply all prescribed methods and instructions every time equipment is employed.

❑ Liquid refrigerants can cause severe frostbite. Avoid possibility of contact through use of adequate gloves and safety glasses.

❑ Refrigerant charge could come from a badly contaminated system. Acid is a product of decomposition; both hydrochloric and hydrofluoric acid can be produced (hydrofluoric acid is the only acid which can etch glass).

Extreme care must be taken to prevent oil spills or refrigerant vapors from contacting skin and clothing surfaces when servicing contaminated equipment.

❑ Wear protective gear, such as safety glasses and shoes, gloves, safety hat or hard hat, long pants, and shirts with long sleeves.

❑ Refrigerant vapors can be harmful if inhaled. Avoid direct ingestion and always provide low-level ventilation.

❑ Assure that all power is disconnected and disabled to any equipment requiring recovery. Lock out any disconnect with an approved locking device.

❑ Never exceed the cylinder's safe liquid weight level, based upon net weight. Maximum capacity of any cylinder is 80 percent by maximum net weight. See "Filling Cylinders" later in this section.

❑ When moving a cylinder, use an appropriate wheeled device. Assure that the cylinder is firmly strapped when using a hand cart. NEVER roll a cylinder on its base or lay it down to roll it from one location to another. Use a fork lift truck for half ton containers of refrigerant recovered from large equipment.

❑ Use top quality hoses. Make sure they are properly and firmly attached. Inspect all hose seals frequently.

❑ Hoses and electrical extension cords can be a trip hazard. Prevent an accident of this sort by placing proper barriers and signs. Place hoses and lines where risk is minimized.

❑ Collect used refrigerant in DOT approved refillable cylinders or drums as appropriate. Approved containers are available from distributors.

❑ Label the cylinder or container as specified in regulations.

❑ If reclaiming, contact the reclaim facility of your choice to arrange transportation.

❑ Assure that all cylinders are in a safe condition, capped as necessary, with proper identification.

Excerpts from "Containers for Recovery of Fluorocarbon Refrigerants," ARI's Guideline K.

Section 5. Responsibility of Owner

This section applies only to ton tanks and cylinders, not drums, because drums are not compressed gas containers (See DOT Title 49 CFR Section 173.301(b)).

5.1 Only the owner may fill proprietary containers or grant permission for some other party to fill them.

5.2 Cylinder/Ton Tank Retesting: **Cylinders and ton tanks must be hydrostatically retested a minimum of once every 5 years in accordance with Title 49 CFR Section 173.34(e) and 173.31(d).** Retesting by visual inspection alone is not permitted.

Responsibility to assure the cylinder or ton tank is within the test date rests with the filler even if the filler is not the owner of the container. If the container is out of date, it shall not be filled. It shall be returned empty to the owner for retest.

Section 6. Labels and Markings
6.1 DOT Requirements

6.1.1 Specific container labeling and marking requirements apply for all DOT regulated hazardous materials. The following instructions apply to non-flammable, recovered refrigerants R-12, R-22, R-500, and R-502 shipped in cylinders and ton tanks.

NOTE: R-11, R-113, and R-114 are not DOT regulated hazardous materials; therefore, DOT labeling and marking requirements do not apply.

USED REFRIGERANT IDENTIFICATION TAG

RECOVERY CONTAINER #: _____

DISTRIBUTOR: _____
(Print Company Name)

OEM: _____
(Print Company Name)

DATE: _____

POSSIBLE CONTAMINANTS:
(What was being cooled? Ex.: water, air, or specify other)

_____

_____

_____

PLEASE LEAVE TAG ATTACHED
EMPTY OR FULL

TO:

FROM:

NON-FLAMMABLE GAS

2

This System is Charged With Du Pont SUVA®

DU PONT **SUVA®** REFRIGERANTS™

MP-39

☐ Use Alkylbenzene Oil **MP-39**
(ex. Zerol® 150T) Type: _____

☐ Desiccant-XH9
or compatible desiccant

☐ Amt. charged _____

Date: _____

This system is charged with Du Pont
SUVA® **MP-39**

H-43858-1

These placards are installed on cylinders used for return of recycled refrigerants.

Alternative refrigerant/oil label for a retrofitted system.

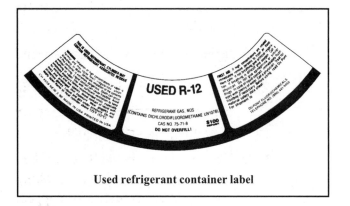

USED R-12

REFRIGERANT GAS, NOS
(CONTAINS DICHLORODIFLUOROMETHANE UN1076)

CAS NO. 75-71-8
$100
DO NOT OVERFILL!

Used refrigerant container label

6.1.2 Labeling: Each cylinder shall display a DOT diamond (square-on-point) "nonflammable gas" label. The 4" X 4" green diamond shaped label may be printed on a tag and securely attached to the cylinder's valve protection cap prior to shipment. Ton tanks require two DOT non-flammable gas labels, one on each end.

6.1.3 Marking: Each container shall be marked with a DOT proper shipping name and an appropriate United Numbering (UN) identification number.

Note: Refer to ARI Guideline K for specific proper names and identification for specific refrigerants.

6.1.3.2 Ton Tanks

UN _____
(insert appropriate UN number as per Section 6.1.3.1)

This marking must appear in 2-inch high letters and numerals on opposite sides of the ton tank.

Cosignee's or nonsignor's name and address must also be shown on each container.

6.3.1 Each container shall display a precautionary label prepared in accordance with American National Standard Z129.1-1982 and Compressed Gas Association Pamphlet C-7 "For Precautional Labels and Marking of Compressed Gas Containers." This label shall include:

- product identity
- antidotes
- signal word
- notes to physicians
- statement of hazards
- instructions in case contact or exposure
- precautionary measures
- instructions in case of fire, spill, or leak
- instructions for container handling and storage

6.3.2 Each container shall display a label stating that it can be used only for a specified recovered refrigerant. Filling instructions shall be provided with the container.

6.3.3 Cylinders and drums shall be marked as shown below in one-inch (minimum) letters and numerals:

RECOVERED   REFRIGERANT
_____ (enter number) This mark shall also appear on the valve end chime of ton tanks.

6.3.4 Printing on labels shall be clear and legible.

6.4 User Information

Each container shall be labeled with the filler's name, address, and date filled.

6.5 Color

The following are examples of coloring schemes for various recovery containers. Depending upon the provider of the recovery container, the actual coloring may vary. However, the use of the color yellow similar to the following examples will identify the container as a recovery vessel.

6.5.1 Cylinders with non-removable collars: **The body shall be gray. The collar shall be yellow.**

6.5.2 Cylinders with removable caps: **The body shall be gray. The shoulder and the cap shall be yellow.**

6.5.3 Drums: **The drum shall be gray. The top head shall be yellow.**

6.5.4 Tons: **The body shall be gray. The ends and chimes shall be yellow.**

Section 7. Filling Procedure

IMPORTANT: DO NOT MIX REFRIGERANT WHEN FILLING CONTAINERS

7.1 Cylinders and Ton Tanks

7.1.1 Do not fill if the present date is more than 5 years past the test date on the container. The test date will be stamped on the shoulder or collar of cylinders and on the valve end chime of ton tanks and appear as follows:

A1
12   89
23

Note: This indicates the cylinder was retested in December of 1989 by retester number A123

7.1.2 Cylinders and ton tanks shall be weighed during filling to ensure user safety. "MAXIMUM GROSS WEIGHT" is indicated on the side of the cylinder or ton tank and shall never be exceeded.

7.1.3 Cylinders and ton tanks shall be checked for leakage prior to shipment. Leaking cylinders and ton tanks must not be shipped.

7.2 Drums

7.2.1 Only recovered refrigerant R-11 shall be placed into a drum that previously contained new refrigerant R-11. Only recovered refrigerant R-113 shall be placed into a drum that previously contained new refrigerant R-113. Drums that originally contained R-113 for use as a cleaning agent shall not be used.

7.2.2 Drums shall be filled to allow a vapor space equal to at least 10 percent of the drum height between the top of the liquid and the bottom of the drum top.

Note: Leaking drums must not be shipped.

Section 8. Transportation

8.1 Local Regulations

The shipper of recovered refrigerant is responsible for determining if there are any state or local regulations restricting transportation, such as classifying recovered refrigerant and oil mixtures as hazardous wastes. The U.S. Environmental Protection Agency does not classify these materials as hazardous waste.

8.2   Shipping Papers: The shipper is required to properly fill out the shipping papers when returning the recovered refrigerant. **The shipping papers must always contain:**

8.2.1   *The quantity and type of container such as:* "2-RETURNABLE CYLINDERS"

8.2.2   The total gross weight of recovered refrigerants.

8.2.3   For DOT hazardous materials, the shipping descriptions must include the following, in sequence:

The DOT proper shipping name, for example: "Chlorodifluoromethane Mixture"

The DOT hazard class, for example: "NON-FLAMMABLE GAS"

The UN identification number, for example: "UN 1018"

8.2.4   For material not regulated by DOT as a hazardous material, the words "Not Regulated by DOT" are recommended, but not required.

8.3 Hazard Identification

When a full or partially full container is shipped, the shipper will be required to affix a DOT hazard label to the container. This is typically a green, 4" square tag reading NON-FLAMMABLE GAS, that can be tied to the valve cover.

If a container is empty and has no residual pressure, a DOT hazard tag is not required.

If the shipper is sending 1,000 pounds (gross weight) or more of a hazardous material on the truck, DOT regulations require the shipper to provide the motor carrier with four non-flammable gas placards. For materials being transported in ton tanks, the placards must also include the appropriate UN four digit identification number. Affixing the placards to the truck is the responsibility of the motor carrier.

For additional information concerning the above regulations, refer to ARI Guideline K.

CFC-HCFC Warning Labels

Containers in which Class I (CFC) or Class II (HCFC) substances are stored or transported in cylinders, must be affixed with a prescribed warning label, in accordance with Section 611 of the Clean Air Act.

Warning labels must also be placed on HVACR equipment containing Class I substances, or on any product manufactured with a process which uses a Class I substance (pipe insulation, packing material, etc.).

It is in violation of regulations if a wholesaler or retailer fails to pass on the warning statement, or if the tag is removed from products purchased or distributed. If the product is repackaged, the label must be replaced.

The statement reads, ***"WARNING: Contains (or manufactured with), a substance which harms public health and environment by destroying ozone in the upper atmosphere."***

Criminal penalties may be imposed for "knowing" (intentional) violations. Civil judgment can be up to $10,000.00 per day per violation. Enforcement of this labeling rule during the first nine months following its publication in the Federal Register, is realized by the EPA to be somewhat difficult due to short notice.

Products containing HCFCs or made using HCFCs, will need to be labeled beginning January 1, 2015. If safe alternatives are available, this ruling on Class II substances could be in effect earlier.

Products manufactured with or containing Class I substances prior to May 15, 1993, or existing in inventory at time of publication of this ruling, are not required to be labeled or tagged.

It is preferred that manufacturers place the warning label on its product or packaging. However, supplemental printed materials, such as warnings on invoices, bills of lading, and MSDSs, could also satisfy the rule, providing they are likely to be read before a purchase is completed.

If a wholesaler assembles components that were manufactured using CFCs but do not contain CFCs, the wholesaler need not pass on warning information unless the wholesaler himself uses any Class I substance in assembling components.

For further information on labeling, call the Stratospheric Ozone Information Hotline: (800) 296-1996 from 10:00 A.M. to 4:00 P.M. EST.

Description of Refillable Refrigerant Cylinders

Refillable cylinders satisfy requirements of DOT 49 CFR 178.51 and 49 CFR 178.61 for specification 4BA and 4BW as shown in the tables below. The 4BA cylinder is composed of two deep-drawn carbon steel heads, welded together with one girth seam. The 4BW cylinder is composed of two separate heads on opposite ends of a center cylindrical section.

Concerns in design pressure of the cylinders are ambient temperature, enclosure of the cylinder within a room, proximity to a heat source, and other factors which may cause excessive pressure due to loss of vapor expansion space in the cylinder.

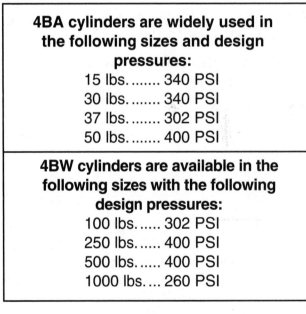

4BA cylinders are widely used in the following sizes and design pressures:
15 lbs. .... 340 PSI
30 lbs. .... 340 PSI
37 lbs. .... 302 PSI
50 lbs. .... 400 PSI

4BW cylinders are available in the following sizes with the following design pressures:
100 lbs. ...... 302 PSI
250 lbs. ...... 400 PSI
500 lbs. ...... 400 PSI
1000 lbs. .... 260 PSI

Filling Cylinders

It is important to be mindful of hydrostatic pressure and possible repercussions of ignoring safe liquid holding capacity of cylinders whenever liquid refrigerant is added to a cylinder.

As heat is added to any cylinder containing liquid refrigerant, vapor pressure increases due to the higher

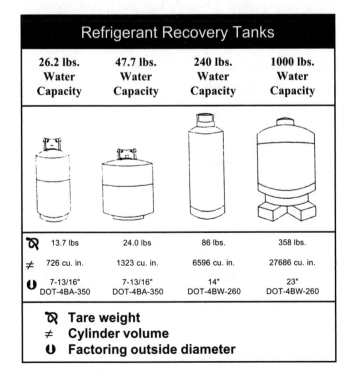

Refrigerant Recovery Tanks			
26.2 lbs. Water Capacity	47.7 lbs. Water Capacity	240 lbs. Water Capacity	1000 lbs. Water Capacity
13.7 lbs	24.0 lbs	86 lbs.	358 lbs.
726 cu. in.	1323 cu. in.	6596 cu. in.	27686 cu. in.
7-13/16" DOT-4BA-350	7-13/16" DOT-4BA-350	14" DOT-4BW-260	23" DOT-4BW-260

℞ Tare weight
≠ Cylinder volume
ʊ Factoring outside diameter

TEMP (F)	PRESSURE		VOLUME		DENSITY	
	PSIA	PSIG	LIQUID	VAPOR	LIQUID	VAPOR
R-12						
100	131.86	117.16	0.012693	0.30794	78.785	3.2474
110	151.11	136.41	0.012924	0.26769	77.376	3.7357
120	172.35	157.65	0.013174	0.23326	75.906	4.2870
130	195.71	181.01	0.013447	0.20364	74.367	4.9107
R-22						
100	210.60	195.91	0.014038	0.25702	71.236	3.8907
110	241.04	226.35	0.014350	0.22222	69.689	4.5000
120	274.60	259.91	0.014694	0.19238	68.054	5.1981
130	311.50	296.80	0.015080	0.16661	66.312	6.0022
R-502						
100	230.89	216.19	0.013895	0.17079	71.967	5.8550
110	262.61	247.91	0.014277	0.14739	70.042	6.7843
120	297.41	282.71	0.014715	0.12711	67.956	7.8669
130	335.54	320.84	0.015229	0.1934	65.661	9.1449

TEMP (F)	PRESSURE (Liquid)		VOLUME		DENSITY	
	PSIA	PSIG	LIQUID	VAPOR	LIQUID	VAPOR
R-402A						
100	226.72	252	0.0154	0.1599	64.97	6.2529
110	304.01	289	0.0161	0.1360	62.14	7.3536
120	345.05	330	0.0169	0.1152	59.02	8.6779
130	390.07	375	0.0180	0.0970	55.53	10.305
R-401A						
100	157.35	143	0.0142	0.3836	70.57	2.6072
110	180.77	166	0.0145	0.3292	68.86	3.0379
120	206.67	192	0.0149	0.2832	67.06	3.5316
130	235.21	221	0.0154	0.2440	65.13	4.0989
HFC-134a						
100	138.996	124	0.0139	0.3408	72.13	2.9347
110	161.227	147	0.0142	0.2912	70.66	3.4337
120	186.023	171	0.0145	0.2494	69.10	4.0089
130	213.574	199	0.0148	0.2139	67.46	4.6745

rate of vaporization. Because liquid and gas are contained, pressure, density, and energy content of the material increases under influence of heat energy. **Hydrostatic pressure occurs when a pressure vessel contains enough liquid to cause tremendous hydraulic force in all directions.** This force is very powerful, and will cause a cylinder to explode if not controlled.

A head of vapor is provided at the top of the cylinder, preventing hydrostatic pressure in case of liquid expansion. Generally, 20 percent of the occupied cylinder volume is devoted to vapor head and considered adequate to prevent hydrostatic pressure from occurring, assuming pure refrigerant is charged into the cylinder.

When refrigerant is reclaimed, it is often diluted with oil which causes specific density values to vary from refrigerant properties tables. This may result in erroneous net weight calculation. Another factor which could disrupt weight calculation is water contamination of the refrigerant.

Designated recovery cylinders are stamped with TW (tare weight) and WC (water capacity) weights, which presents a choice of refilling methods. Stamped with gross weight values, large bulk containers allow refrigerant filling procedures to be a straight-forward process.

To safely assure adequate vapor head within a cylinder about to receive liquid refrigerant, the correct amount of liquid should be weighed in with a calibrated scale. Never estimate how much refrigerant is in a cylinder.

Standard Method of Cylinder Filling

The following is one method that can be employed to safely and accurately add refrigerant to an approved cylinder:

1. Assume highest expected ambient temperature, generally accepted as 130°F.

2. Determine TARE weight of the cylinder in pounds. If this is not stamped somewhere on cylinder shoulder, weigh empty cylinder.

3. Determine cylinder volume in cubic feet by consulting cylinder manufacturer.

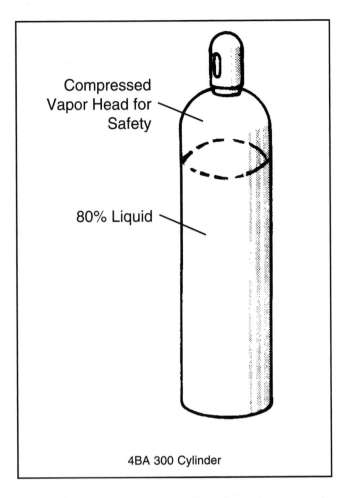

Compressed Vapor Head for Safety

80% Liquid

4BA 300 Cylinder

Alternative Method of Cylinder Filling

An alternative method to determine safe holding capacity is in reference to stated water capacity (WC) on a refrigerant cylinder. Multiply the water capacity times the filling density of the refrigerant. The factor of filling density is found by taking the refrigerant saturated liquid density at 130°F and dividing it by the density of water (62.4 .b/cu.ft.). See example illustrated on page 116.

Tare weight of the cylinder must be determined through the manufacturer, or the cylinder weighed in an empty state as new. The "empty" cylinder must be weighed to determine if it contains oil. If oil or other substances are proven to be present, apply low nitrogen pressure to the cylinder. Invert it if necessary, and carefully discharge contents into a waste container. Check cylinder true empty weight on a frequent basis.

Caution: This filling method conforms to DOT Specification 49 CFR. However, if there is possibility of ambient temperature exceeding 130°F at any time during storage of a cylinder filled per direction according to the "Alternative Method" there could be a sudden release of refrigerant from the cylinder. To assure safety, multiply calculated weight by .80.

Refrigerant recovery equipment is available which automatically shuts off when its mounted recovery cylinder has attained 80 percent of liquid capacity by full cylinder weight by volume. (Refer to Recovery Tank 80% Stop/Fill Switch example on page 116.) **Repeated use of this cylinder could cause an erroneous indication of weight and faulty shut-off point.**

Refilling Precautions

Before refrigerant recovery procedure or when it may become necessary to transfer refrigerant from one container to another, pertinent precautions must be taken in order to assure safe recovery or transfer.

Almost all cylinders used for refrigerant are of the deposit type. The xception is any disposable style

4. Using the Saturated Tables (on page 114) appropriate for refrigerant to be added, determine LIQUID specific density (lb./cu. ft.) at highest anticipated ambient temperature.

5. Multiply internal cylinder volume (cu. ft.) by determined liquid specific density (lb./cu. ft.) to determine the full capacity of the cylinder in weight pounds (lbs).

6. Multiply full weight by .80 to establish safety factor of 20 percent vapor head.

7. With the cylinder on a scale, weigh in only that amount which equates to 80 percent liquid capacity as NET weight-lbs.

   TARE weight + NET weight = GROSS weight in pounds.

Alternative Cylinder Filling Method

EXAMPLE:

Density of refrigerant at 130°:				Density of water:		Filling density:
R-12	=	74.36 lb./cu. ft.	÷	62.4 lb./cu. ft.	=	1.19
R-22	=	66.312 lb./cu. ft.	÷	62.4 lb./cu. ft.	=	1.06
R-500	=	65.090 lb./cu. ft.	÷	62.4 lb./cu. ft.	=	1.04
R-502	=	65.661 lb./cu. ft.	÷	62.4 lb./cu. ft.	=	1.05

For this example, we are to assume a 50 lb. recovery cylinder, stamped with a water weight capacity of 47.6 lbs. Now, determine cylinder safe liquid fill capacity.

R-12	47.6 lbs. water weight	X	1.19	=	56.60 pounds (net weight)
R-22	47.6 lbs. water weight	X	1.06	=	50.46 pounds (net weight)
R-500	47.6 lbs. water weight	X	1.04	=	49.50 pounds (net weight)
R-502	47.6 lbs. water weight	X	1.05	=	49.98 pounds (net weight)

Recovery Tank 80% Stop-Fill Switch

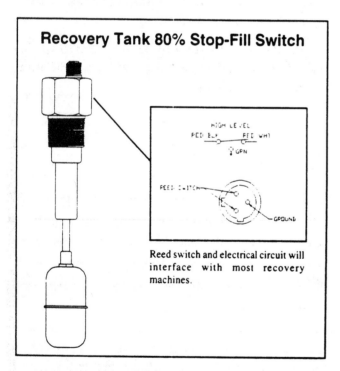

Reed switch and electrical circuit will interface with most recovery machines.

container. Ownership of the container is usually retained by the manufacturer of the refrigerant. Use for any purpose other than removal of original refrigerant is prohibited. In order to use these cylinders for another purpose, a formal waiver from the manufacturer is required. Storage of used or contaminated refrigerant, or use for recovering, reclaiming or recycling of refrigerant is inclusive of a waiver. **Disposable refrigerant cylinders are to be used for shipping the original refrigerant only. Any further use is prohibited.**

When a proper cylinder is selected for refilling, always check its integrity. Inspect it visually for rust, abrasion, and dents, and examine the valve assembly for damage or tampering. Check the expiration date; it must be less than five years since last date of certification. Make sure the cylinder is of proper color and the identification labels conform to regulations, with proper refrigerant indicated.

Never mix refrigerants. Once a certain refrigerant is mixed with another, the expensive process of distillation will be required or as a last resort incineration will be required.

ALWAYS weigh in liquid, allowing for at least 20 percent vapor head for expansion of liquid. Be sure to check NET vs. GROSS weight.

Vehicular Transportation

When transporting any pressurized cylinder or drum, no matter what size, extreme care must be taken to avoid serious injury to the person transporting, and danger to innocent individuals. If disposable cylinders are allowed to collide with paint scraping impact, serviceable life of the cylinder is shortened at the very least. If the blow is sharp enough, physical damage to the cylinder could occur. Either way, refrigerant is sure to eventually escape.

New large cylinders are shipped on pallets, with pre-cutout spacers to hold them solidly. Request one from your supplier and secure it to your truck. Provide a solid and safe means to connect a chain or stout cable, securing one end and providing a means for tight connection for the other. **A cable or chain holds the cylinder(s) rigid in an upright position, as required by DOT.**

Apply the same method for those smaller cylinders, which can be awkward to place. Produce a smaller device to hold them. Remember that any time those cylinders get skinned, they may rust and the walls can become weakened just from the blow.

Safe Disposal Requirements

Under the EPA's rule, equipment that is typically dismantled on site before disposal (e.g., retail food refrigeration, cold storage warehouse refrigeration, chillers, and industrial process refrigeration), has to have the refrigerant removed and recovered in accordance with the EPA's requirements for servicing. However, equipment that typically enters the waste stream with the charge intact is subject to special safe disposal requirements (e.g., motor vehicle air conditioners, household refrigerators and freezers, and room air conditioners). Refer to Section Six.

Under these requirements, **the final person in the disposal chain is responsible for ensuring that refrigerant is recovered from equipment before final disposal of the equipment** (e.g., a scrap metal recycler or landfill owner). However, persons "upstream" could remove the refrigerant and provide documentation of its removal to the final person if this is more cost effective.

Technician certification is not required for individuals removing refrigerant from appliances in the waste stream.

Notes

Notes

SECTION TEN
Conservation - Servicing and Testing

Introduction of recovery, recycling, and reclamation of CFCs, HFCs, and HCFCs in recent years has caused changes in decades-old service techniques and procedures. Gone are the days when a few ounces or pounds of refrigerant exhausted to the atmosphere drew little concern. Now these changes are of serious concern to everyone.

With increasing refrigerant costs, production limitations, and added costs for recovery/recycling equipment, the most practical and efficient service methods must be practiced. Correct installation practices and precise leak detection means are also important.

The Prohibition on Venting

Effective July 1, 1992, Section 608 of the Clean Air Act prohibits individuals from knowingly venting ozone-depleting compounds used as refrigerants into the atmosphere while maintaining, servicing, repairing, or disposing of air conditioning or refrigeration equipment.

The EPA recognizes that some common maintenance and repair procedures that are not associated with efforts to recover or recycle may release a small quantity of refrigerant. Such releases constitute violations of the prohibition on venting. The Agency, however, will consider the circumstances of a refrigerant release determining whether or not to pursue an enforcement action.

Service Aperture

All air conditioning and refrigeration equipment, except for small appliances and room air conditioners must be provided with a servicing aperture that facilitates recovery of the refrigerant.

Process Stub

The EPA is requiring that small appliances and room air conditioners be provided with a process stub to facilitate removal of the refrigerant at servicing. **The Agency defines "process stub" as a length of tubing that provides access to the refrigerant inside a small appliance or room air conditioner, and that can be resealed at the conclusion of repair service.**

Schrader Valves

The Agency is not prohibiting the use of schrader valves on small appliances. It is believed that such valves assist in the recovery of refrigerant, and concerns for their release of refrigerant can be minimized through proper use. All schrader valves should be capped while not in use. The Agency is not dictating where the servicing aperture or process stub should be placed.

Service Practice Requirements

Technicians are required to evacuate air conditioning and refrigeration equipment to established vacuum levels. **Refer to Section Eight for detailed information.**

Limited Exceptions

Technicians repairing small appliances, such as household refrigerators, household freezers, and water coolers, are required to recover **90 percent of the refrigerant in a system with a functioning compressor, and 80 percent of the refrigerant in a system with a non-operating compressor.** (Refer to Section Seven, Table 7-1.) "Small appliance" means products that are fully manufactured, charged, and hermetically sealed in a factory, **with five (5) pounds or less of refrigerant.** Such products include window air conditioners and packaged terminal heat pumps, dehumidifiers, under-the-counter ice makers, vending machines, and drinking water coolers. On the other hand, **equipment containing more than 200 lbs. of charge would be subject to a standard that requires recovery of over 99 percent of the refrigerant.**

The EPA has established limited exceptions to its evacuation requirements for repairs to leaky equipment and for repairs that are not major and are not followed by an evacuation of the equipment to the environment.

"Major" repairs include those involving removal of the compressor, condenser, evaporator, or auxiliary heat exchanger coil.

"Minor" repairs involve an operation on equipment which is not followed by an evacuation to the environment (e.g., addition of refrigerant through a process port, replacement of pressure switches, etc.).

Leaks that make required evacuation levels impossible to obtain also allow exceptions to evacuation requirements. (See "Refrigerant Leaks" on page 126.)

If, due to leaks, evacuation to a specified level is not attainable, or would substantially contaminate the refrigerant being recovered, persons opening the appliance must:

❑ Isolate leaking from non-leaking components wherever possible.

❑ Evacuate non-leaking components to EPA specified vacuum levels that can be attained without substantially contaminating the refrigerant. This level cannot exceed 0 psig.

If evacuation of the equipment is not to be performed when repairs are complete, and if the repair is not major, then the appliance must:

❑ Be evacuated to at least 0 psig before it is opened if it is a high (or a very high) pressure appliance.

❑ Be pressurized to 0 psig before it is opened if it is a low pressure appliance.

The EPA has also established that refrigerant recovered and/or recycled could be returned to the same system or other systems owned by the same person without restriction. If refrigerant changes ownership, however, that refrigerant must be reclaimed (i.e., cleaned to the ARI 700 standard of purity and chemically analyzed to verify that it meets this standard).

System Dependent Equipment (Passive)
When system compressors are operable but are not capable of achieving required vacuums with a single recovery unit (a situation which the EPA believes will be relatively rare), the required vacuum can be achieved by using a second, evacuated cylinder. **Thus, use of system-dependent equipment that has been certified under ARI 740 with appliances containing up to 15 pounds of refrigerant is permitted.** Because system-dependent equipment is not able to achieve required vacuums if the system compressor is not operating, **the EPA is requiring technicians who repair or dispose of appliances other than small appliances to have at least one self-contained recovery device.** The recovery unit should be available at the shop to recover the refrigerant from systems with non-operating compressors.

Vacuum Pump Guidelines

After recovery and service of the equipment, evacuation using a vacuum pump is suggested for dehydration of a refrigeration or air conditioning system. Moisture left in the system will cause acids to form. **During evacuation, the system's vacuum gauge should be located near the system's tubing and as far from the vacuum pump as possible.** For any given system, the vacuum pump capacity in CFM and the vacuum pump's suction line diameter will determine how long it will take for the system to be dehydrated. Always measure the final vacuum with the vacuum pump isolated from the system and turned off. **If the system is to be recharged with refrigerant, never heat the refrigerant storage**

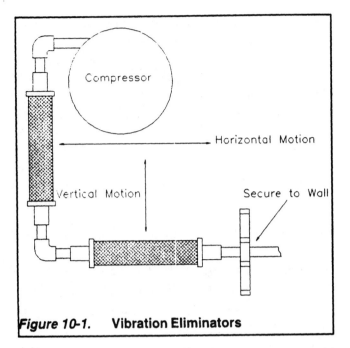

Figure 10-1. Vibration Eliminators

tank or the recovery tank with an open flame. Instead, partly submerge the tank in warm water to bring up the tank pressure. Heating the tank with an open flame may cause the refrigeratant in the tank to decompose forming harmful by-products.

If a technician uses a vacuum pump that is too large for the system, the moisture in the system may freeze due to the quick drop in pressure. Dry nitrogen may then be introduced to the system to bring up the pressure and avoid freezing of the moisture. Also, if the system is under a deep vacuum, the compressor should never be energized. This will prevent motor winding damage from occurring due to a short circuit of the motor terminals inside the compressor.

Installation Recommendations

Proper installation methods can prevent refrigerant leakage. Good design criteria is of primary importance. Properly installed quality fittings, with a high standard of craftsmanship and keen attention to detail will accomplish the goal of a leak-free system.

Strategically placed shut off valves, and perhaps receivers to facilitate minimal refrigerant loss from a system when it does need service, must be used not only when a system is installed but also as retrofit when systems are serviced. Allow all components to be segregated and system refrigerant to be contained, except that which must be recovered.

Proper placement of suspension brackets to alleviate stress points due to thermal expansion of tubing is critical in preventing serious future system leakage due to line breakage. Caution in installation of 45° ells with consideration of mechanical stress on the fitting and installation of effective vibration eliminators in lines, are examples of measures which must be taken (see Figure 10-1 and Figure 10-2).

Soldering and Brazing Technique

Soldering and brazing quality is a very important part of the installation. Soldering skill can be learned only through understanding basic technique and many hours of practice. Brazing is similar, but potentially more dangerous due to necessary equipment. It is not the intention here to teach these skills, but to provide a check list of very important points to use as a guideline.

❏ Thoroughly think the job out before approaching it. Make sure all necessary component parts of the assembly are of the proper types and laid out for identification to the print.

❏ Tubing must be measured accurately to assure complete insertion into any fitting.

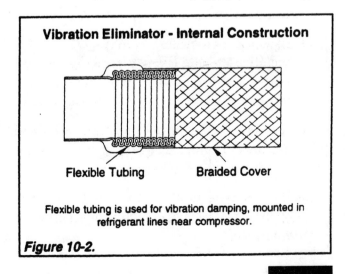

Flexible tubing is used for vibration damping, mounted in refrigerant lines near compressor.

Figure 10-2.

- Cuts must be made squarely, thoroughly reamed, outside edges filed if necessary (for smooth insertion), and polished with sand cloth as required.

- Tubing ends must be clean and free from any oils; even from the fingers.

- Hand fit all tubes with fittings in temporary fashion before soldering or brazing. After final adjustments are made, most of the assembly can be brazed with one operation. Suspend tubing by the brackets to be permanently installed, fitting all sections to required design specification for slope and run.

- Employ only the proper flux for whatever soldering/brazing material that is being used. Do not substitute.

- Assure that material employed is of the proper grade and alloy. Hard silver solder is always best for steel to brass or copper, and copper to brass or brass to brass.

- Apply flux properly. The flux does not "glue" the fitting together—use it sparingly; flux spreads widely as heat is added. Its purpose is to prevent oxidation while the joint is brazed or soldered. Use a minimum amount applied to short end of the male fitting to prevent acid from entering the system.

- After assembling the joint, apply heat to the heaviest portion first. Initially distribute temperature evenly, then concentrate heat into the joint at the base to draw solder deeply into the fitting.

- Apply only as much heat as necessary; do not overheat the fitting. Too much heat will drive the solder or brazing material out of the gap, perhaps even into the line.

- The temperature of the fitting and line should cause the filler material to melt, not the direct heat from the torch.

- Keep the torch moving, as the flame is the tool which draws the filler like a magnet.

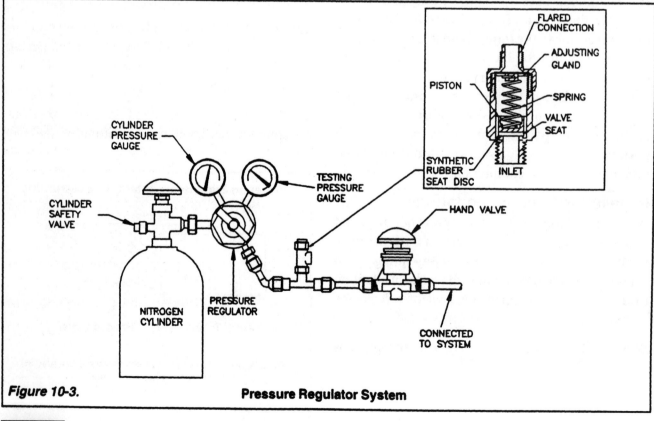

Figure 10-3.   **Pressure Regulator System**

Refrigerant Transition and Recovery Certification Program Manual

□ Heaviest concentration of heat energy is at the tip of the brightest conical flame (near torch tip). The very point of this bright flame is the absolute hottest.

□ Vary the intensity of heat as required through shortening and lengthening your approach to the fitting with the torch.

REMEMBER: Use a minimum amount of flux applied in the right place, a minimum amount of heat applied in a metered manner, and a minimum amount of filler material to completely fill the void in depth.

ALSO: Use proper filler material. Some products do not require flux at all—if so, do not use flux. Others require a specified kind—do not substitute.

It is highly recommended that dry nitrogen be used when soldering or brazing copper fittings and tubing. If an inert gas is not employed to assure an oxygen-free atmosphere within the copper tubing while applying heat, copper oxide will be produced within the copper tubing as it cools. Think of copper oxide as "copper rust."

A nitrogen cylinder containing a full charge possesses approximately 2,500 PSI. A pressure regulator must be used. An adjustment to produce a very slow flow is all that is needed to displace oxygen within the tubes. Refer to page 122.

CAUTION: Always make sure there is a safety pressure relief valve installed on the down-stream side of the regulator with a release pressure setting of no greater than 150 psi. The regulator could fail, and a disaster could result (see Figure 10-3).

Leak Testing

After system charge has been recovered, pressurize with nitrogen to a pressure no greater than design low side test pressure. Leak check with specially formulated bubble solution.

Note: Some equipment safety relief valves have a setting as low as 200 PSI. Never approach the release pressure of the relief valves; once they have released, they must be rebuilt and calibrated.

Systematically leak check any possible place where refrigerant could escape. Do not limit yourself to only that portion of the system soldered or brazed.

The following leak check devices and methods have historically good results:

1. Soap Bubble Testing is inexpensive and good for finding and pinpointing leaks. Bubble solutions are commercially available specifically for leak testing. This method can be used with a nitrogen-charged system.

2. Electronic Halide Leak Detectors use an ionization cell to detect presence of halides. These detectors are effective at detecting small leaks of as low as 1/2 oz. per year. When using leak detectors without a varying sensitivity, a large leak may saturate the element.

   Some have adjustable sensitivity ranges with a variable "tuning" means for sensitivity. An experienced person can employ this form of leak detection with very good results.

3. Ultrasonic Leak Detectors employ an ultrasonic method to "listen" for leaking gas. This requires some advance knowledge of the leak location and a fairly low background noise level. They do provide good sensitivity adjustment.

   Ultrasonic leak detectors are reliable for outdoor use where air currents can upset accuracy with other methods of detection.

4. Halide Torch Detector depends upon a flame changing color in response to refrigerant pulled into it through a flexible tube. It requires some skill in discernment, and has little use outdoors in bright sunlight. It is not recommended where there is danger of explosive fumes or gases.

5. **Standing Pressure** contained by the system to be tested is effective if system internal volumes are not large. This is generally done after all leaks have been found and corrected and the system has been leak checked again. The system is pressurized to 125 psi with nitrogen and allowed to sit over night for later observation.

6. **Standing Vacuum Test** is a viable means, if the system does not have possibility of water entering the evaporator (water chiller) or condenser (water-cooled type). **A deep vacuum pump is necessary in order to achieve a standing vacuum of at least 500 microns.** The objective of the deep vacuum is primarily a check to assure system dehydration and degassing. This method is usually implemented if equipment is of the class applied to residential and light commercial.

A deep vacuum is attained and the unit valved off to observe any minute loss of vacuum after system equalization. Severity of loss over a given period of time would indicate either presence of a leak or moisture within the system. The procedure is monitored through a micron gauge to sense system vacuum conditions.

7. **Electronic Leak Detector — R-22 trace gas mixed with nitrogen can be used for leak checking. This mixture can be released into the atmospheric.**

NEVER employ standard compressed air for leak checking purposes. As air is compressed through a normal compressor for that purpose, compressor crankcase oil is in the air as vapor. This oil is certainly contaminated and it contains wax. There is also risk that water vapor is contained in the air.

If the refrigeration system becomes contaminated with impure oil (other than pure grade refrigeration oil) or wax-producing substances (such as solder flux), floc point will certainly occur when the machine is put into operation.

CAUTION: Air contains 18 percent oxygen, which when mixed with many refrigerants (as would be done in a leak-checking procedure), could cause an explosive mixture if flame is encountered. Never leak check with pressurized air and refrigerants.

Leak Detectors for Alternative Refrigerants

Introduction of alternative refrigerants has required development of new leak detection equipment (see Table 1). Most of the older technology leak detectors were designed to monitor for presence of chlorine in a refrigerant molecule. HFC alternative refrigerants have no chlorine. Consequently, leak detectors designed for CFCs and HCFCs are either unable to detect some of the new hydrofluorocarbon (HFC) alternatives at any level, or are unable to detect them at a level low enough to meet user's requirements.

Both HFC-134a and HCFC-123 are alternative refrigerants that have required redesigned leak detection equipment. For HFC-134a, a detector had to be designed that could detect a fluorine compound. HCFC-123 requires the sensitivity to be increased by a factor of ten. The primary need for a HFC-134a detector is in pinpointing a leak in a refrigeration system. For HCFC-123, a leak detector capable of monitoring the ambient refrigerant level in an equipment room is needed.

Pinpointing leak detectors are typically hand-held units. They are normally used to find small leaks that cannot be located visually or by sound. Most units can detect leaks of 0.5 oz./yr. (approximately 50 ppm). The majority of pinpoint detectors available today are battery powered; thus, they are portable.

Use of an area monitoring detector to indicate presence of refrigerant is analogous to using a smoke detector at home to detect the presence of a fire. These detectors cannot determine the location of a leak, but can determine existence of a leak. The actual source of a leak must be determined with perhaps a pinpointing detector, as previously described.

There are six main types of detectors that can be used for monitoring alternative refrigerant leaks:

1. **Non-selective detector**
2. **Halogen specific**
3. **Compound specific**
4. **Infrared-based**
5. **Fluorescent dyes**
6. **Ultrasonic**

Non-Selective Detectors, which include metal oxide semiconductors and thermal and/or electrical ionization, are the least expensive, costing between $200.00 and $1,400.00. They are generally simple to use and quite durable; however, they have the poorest detection limit (typically 50 ppm) and can be activated by other compounds.

Most of these units cannot be calibrated to give an exact concentration of refrigerant in an area. The

Table 1

Comparison of Various Types of Leak Detectors			
	Non-Selective	**Halogen-Selective**	**Compound-Specific**	**Fluorescent Dyes**

	Non-Selective	**Halogen-Selective**	**Compound-Specific**	**Fluorescent Dyes**
Advantages	• Price ($250-$1,500) • Simplicity • Ruggedness	• Simple/rugged • Can be calibrated • Good sensitivity • Low maintenance	• Virtually interference-free • Can be calibrated • Good sensitivity	• Low cost • Little specialized equipment required • Good detection limits • Rapid detection • Interference-free
Disadvantages	• Poor detection limits • Cross-sensitive to other species • Most cannot be calibrated	• Price ($280-$2,500) • Not compound specific • Detector lifetime/ stability	• Price (~$10,000; lower though chiller manufacturers); IR-PAS may be priced lower • Complexity/ maintenance • Stability?	• Potential compatibility problems • Cannot be automated • Cannot be calibrated • Some areas not observable
Vendors	• >6 for leak pinpointers • 3-4 for area monitors	• 3 for leak pinpointers • 2 for area monitors	• Several for IR • Several working on IR-PAS	• 2 currently exist
Applications *Leak Pinpointing*	• HFC-134a, HFC-123, Blends	• All HCFCs, HFC-134a, Blends	• Not recommended due to high price	• Works with most systems
Area Monitoring	• Not recommended due to cross-sensitivity and poor detection limits	• All HCFCs, HFC-134a, Blends	• All HCFCs, HFC-134a, Blends	• Not applicable
Other	• None	• Where only one refrigerant is used • In moderately clean equipment rooms	• "Dirty" environments • Multi-refrigerant environments	• None

primary use for non-selective detectors is pinpointing the source of a leak. Lack of sensitivity, inability to be calibrated, and cross-sensitivity to other compounds make non-selective detectors poor candidates for area monitoring applications.

Halogen-Specific Detectors use a specialized sensor that allows the monitor to detect compounds containing chlorine, fluorine, and bromine, without being activated by other species. The major advantage of halogen-specificity is a reduction in the number of "nuisance alarms" (false alarms) caused by the presence of some compound in the area other than a refrigerant.

They are easy to use, feature higher sensitivity (typically 5 ppm) than the non-selective detectors, and are quite durable. In addition, they can be calibrated. Current costs for halogen-specific detectors are approximately $300 for pinpointing-type and $1,800 for an area monitor.

They are best suited for use in moderately clean equipment rooms where only one refrigerant is being used. Lack of response to non-halogenated compounds also allows them to be used in storage areas or areas where other (non-halogenated) compounds may be present.

Compound-Specific Detectors are the most expensive of the leak detectors, costing between $6,000 and $20,000. Compound-specific leak detectors are virtually interference free and offer the highest selectivity of all others currently available. These complex detectors can be calibrated and use a variety of technologies, such as infrared spectroscopy (IR), ion-mobility spectrometry (IMS), photoacoustic spectroscopy (PAS), and gas chromatography (GC).

Infrared-based instruments have been in use for refrigerant leak detectors for some time now, and are proven performers. The IMS- based instruments can detect alternative refrigerants at very low levels (1 ppm); however, current systems are expensive (approx. $12,000).

Ideal for use in "dirty" environments and those with multiple refrigerants, the IR-based detectors are effective for area monitoring of HFCs, HCFCs, and blends. However, due to the high costs involved, these detectors are not recommended for pinpointing leaks.

Fluorescent dye leak detection requires addition of a measured amount of substance to the compressor crankcase oil. Mixed with system oil charge, substance circulates throughout the circuit when unit is running. At a leak location, escaping gases move an amount of the additive to exterior tubing surfaces. When an ultraviolet lamp is directed to this location, an obvious glowing effect is produced, pin-pointing the leak.

The advantage is low cost, with little specialized equipment needed. Noteworthy is interference-free rapid detection. One disadvantage is possible leakage from an inaccessible and awkward area. Calibration is not feasible.

Ultrasonic leak detection is ideal for leakage of most gases. It is a hand-held unit which can be calibrated. This leak detector senses characteristic sound waves of escaping gases emitted from a puncture or small orifice. Sensitive to sound beyond the range and amplitude of human hearing, this device indicates when such sound is emitted due to a leak.

The advantage is that regardless of refrigerant charge, the device will effectively detect a leak. Possible disadvantage is interference by sound from ambient sources disrupting accurate leak detection. Protective shrouds are available to place over the tip, shielding extraneous ambient noise.

Refrigerant Leaks

EPA rules address refrigerant releases that take place during servicing and disposal of air conditioning and refrigeration equipment, because such releases account for between 50 and 95 percent of total emissions over the lifetime of this equipment.

Owners of equipment with refrigerant charges of greater than 50 pounds weight are required to repair substantial leaks. Substantial leaks are:

- A 35 percent annual leak rate is established for the industrial process and commercial refrigeration sectors as the trigger for requiring repairs.

- A second annual leak rate of 15 percent of charge per year is established for comfort cooling chillers and all other equipment with a charge of over 50 pounds weight other than industrial process and commercial refrigeration equipment.

Within 30 days of discovery of a leak, the equipment owner must authorize repair of substantial leaks or develop an equipment retirement/retrofit plan. The plan shall detail activities planned to retire equipment and replace it with equipment that uses a non ozone-depleting or a less ozone-depleting substance or to retrofit existing equipment within one year.

Owners of air conditioning and refrigeration equipment with more than 50 pounds weight of charge must keep records of the quantity of refrigerant added to their equipment during servicing and maintenance procedures. These records must also include refrigerant purchased and added to equipment each month in cases where owners add their own refrigerant. The technician should inform the owner when a leak exists.

Service-Related Suggestions

The following service-related suggestions are applicable:

- Maintain accurate logs of refrigerant used in large systems or in cases of service contracts. Refer to Appendix III.

- Periodically inspect systems, looking specifically for signs of leakage. A visual inspection can expose traces of oil which would indicate a refrigerant leak.

- Purchase good quality leak detection equipment with ability for sensitivity adjustment. Develop skill in usage.

To minimize waste:

- Discontinue past wasteful and careless uses of refrigerant.

- On a frequent and regular basis, check testing equipment for flaw, such as charging hose integrity and manifold valve closure tightness. The rubber seal on charging hoses can fail after only a few uses.

- Select and purchase low leak rate, non-permeable hoses of highest quality. This will also aid in evacuation procedures.

- Minimize loss of refrigerant when purging air from hoses prior to installation. It does not take much purge to expel that small internal volume of the hose.

- In commercial and central air conditioning systems, include as many shut off valves as can be installed to segregate major components. Segregated component areas may require pressure-electrical safety devices to provide adequate protection against over-pressure operating conditions.

There is not a refrigerant produced which adversely affects the environment, as long as it remains within the system.

CFC Demand

It is estimated that 30 percent of all CFC refrigerant used comes from reclamation. With new production of CFCs halted and virgin material not available, many units will have to be replaced at high cost to users. Leaks, emissions, and burnouts will present a natural attrition. Many newly installed and operating refrigeration and air conditioning systems have a service life expectancy of 15 to 30 years. Availability of refrigerant for service or repair will come out of the machines themselves.

Economic sense will mandate conservation, with increasing costs and taxes on refrigerants. Ability to expertly maintain and service equipment will depend upon the service personnel's knowledge and craftsmanship.

Notes

SECTION ELEVEN
Chillers

- ❑ Low Pressure Centrifugal Chillers
- ❑ Replacement Refrigerants
- ❑ Safety and Handling Procedures
- ❑ Common Safety Related Questions
- ❑ Refrigerant Decomposition
- ❑ Equipment Room/Job Site Requirements
- ❑ Safety Group Classification System
- ❑ Retrofitting of Chillers with Alternative Refrigerants
- ❑ R-123 Performance
- ❑ R-123 Compatibility and Handling
- ❑ Retrofit of R-134a into Existing Equipment
- ❑ Oil/Refrigerant Relationships
- ❑ R-12/R-134a Procedure for Retrofit
- ❑ Planning and Acting for the Future

Low Pressure Chillers

Low pressure chillers are used for large installations and range up to 2,000 tons in a single unit. Low pressure chillers typically use R-11, R-113, or R-123 as their refrigerants. Larger units over 150 tons of capacity will use R-11 while smaller units most often use R-113. A HCFC replacement refrigerant R-123 is an acceptable substitute for CFC 11 (R-11) under the Significant New Alternative Policy (SNAP) program.

Other replacement refrigerants including R-123 will be analyzed within this section.

Centrifugal System Configuration

Compressor

The operating principle of a centrifugal refrigerating compressor is very similar to that of a centrifugal liquid pump or centrifugal air moving fan. (See Figure 11-1.)

Through centrifugal force developed by the impeller wheel, vapor is forced radially outward between narrowing walls of the hollow impeller wheel, causing an increased refrigerant density and raised

internal kinetic energy through self compression. Refrigerant is rapidly discharged from the blade tips into a housing within the compressor body at an increased density, temperature and pressure.

Lubrication

Some centrifugal compressors are pressure lubricated, typically with the oil sump and pump as part of the compressor body. Others such as the hermetically driven have refrigerant-cooled motors, while open drive units depend on air to cool the motor.

Condenser

Condenser design is of the shell and tube type with water being the condensing medium. Through multiple water passes, refrigerant vapor is condensed within the shell. Condenser water is usually pumped to a cooling tower where atmospheric air is passed over the water to remove the heat, allowing most of the water to be reused.

Float Type Metering Device

Two types of floats are used, the high-side float and the low-side float. The high-side float regulates refrigerant through the high-side float valve to an intermediate chamber between the high-side float chamber and the low-side float chamber. Some vapor is directed to the second stage impeller suction inlet while the remaining cooled refrigerant leaves the intermediate chamber into the low-side float chamber.

From the low-side float chamber, liquid refrigerant is metered into the evaporator chiller shell where heat is transferred from the circulated load.

Evaporator or Cooler

The typical evaporator assembly is of the shell and tube flooded type with refrigerant in the evaporator shell. Secondary refrigerant (water) is circulated

129

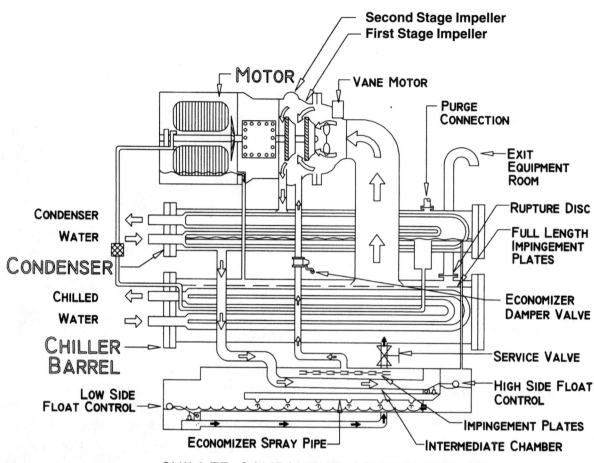

Figure 11-1.

CHILLER COMPONENT CONFIGURATION

within a closed loop through tubes and pumped out to the heat load or air handler. **Pressure relief is accomplished by rupture discs, located on top of the evaporator or cooler, set to fail at 15 psig. (See Figure 11-1.)**

Capacity Control

System capacity control is accomplished at the compressor by one or a combination of the following methods:

- ❑ Suction throttling damper
- ❑ Hot gas bypass
- ❑ Variable pitch inlet guide vane
- ❑ Variable speed driver (turbine)
- ❑ Variable-frequency drive

Purge Systems

Air and non-condensables tend to rise and collect at the highest point in a system. **That location is always within the top section of the shell and tube condenser, with the condenser located at the highest location of a centrifugal appliance. (See Figure 11-2.)**

Because low pressure systems operate in a vacuum, non-condensables can enter the system through leaks in fittings and gaskets. Auto purge units are used to purge and vent these non-condensables from the top of the system's condenser. The pure refrigerant is then returned to the evaporator section. **While the EPA allows purge systems to release small amounts of refrigerant along with the non-condensables, a purge system that has an increase in running time is a good indication that the refrigeration system has a leak and is taking in**

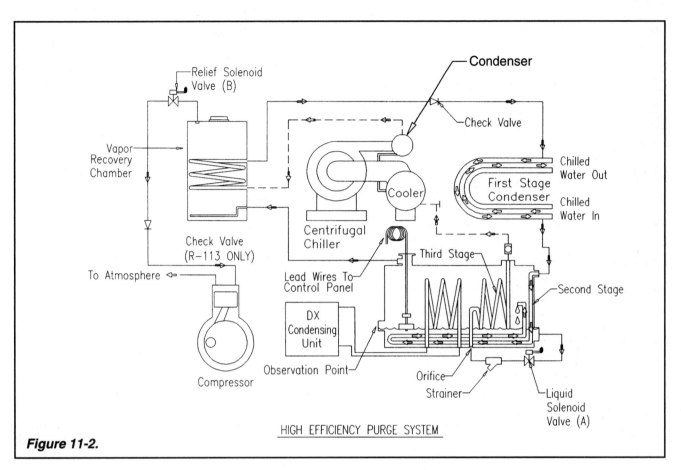

Figure 11-2.

HIGH EFFICIENCY PURGE SYSTEM

Labels in figure:
- Relief Solenoid Valve (B)
- Vapor Recovery Chamber
- Check Valve (R-113 ONLY)
- To Atmosphere
- Compressor
- Condenser
- Check Valve
- Chilled Water Out
- First Stage Condenser
- Chilled Water In
- Cooler
- Centrifugal Chiller
- Lead Wires To Control Panel
- DX Condensing Unit
- Observation Point
- Third Stage
- Second Stage
- Orifice
- Strainer
- Liquid Solenoid Valve (A)

non-condensables. **If the leak is substantial, the EPA requires its repair.** Newer, high efficiency purge systems are available that reduce refrigerant emissions by 97%.

Leak Checking

Leak checking low pressure systems is done with refrigerant system pressure. Heater blankets or the circulation of hot water will raise system pressures and a conventional leak detection method can then be used. **Pressures in the system should never exceed 10 psig because the rupture disc on the evaporator is set to fail at 15 psig. The rupture disc's discharge line should always be piped to the outdoors. When checking a low pressure system for tube leaks, a hydrostatic test tube kit can be used.** Remove the water and place the leak detector probe at the drain valve to detect leaks.

During a standard vacuum test, if the pressure in the system rises from 1mm of Hg to 2.5mm of Hg, the

system probably has a leak and should be leak checked.

Refrigerant Recovery

Recovery machines have been employed with large centrifugal equipment for many years when there is a need for major service. Refrigerant recovered from a low pressure chiller begins with liquid removal followed by vapor removal. Often, recovery will cause local cold spots to occur in the system. **To prevent water from freezing in the system, make sure the system's water pumps, recovery compressor, and recovery condenser water is turned on. Recovery condenser water usually comes from the local water supply**

The low pressure chiller's recovery unit has a pressure limiting device that should be set for a cut-out of 10 psig. This is because the rupture disc on a recovery vessel for low pressure refrigerants is set to fail at 15 psig.

Low pressure appliances employ a flooded type evaporator called a "chiller barrel" (Figure 11-1). Due to convenient vapor and liquid access from a single location in the system, low side, centrifugal low pressure appliances are refrigerant-side serviced through the access valves located at the chiller barrel only.

Liquid Recovery

Liquid recovery is done by using the push-pull method as show in Figure 11-3. A lower pressure is created in the recovery cylinder which forces the liquid refrigerant from the chiller into the cylinder.

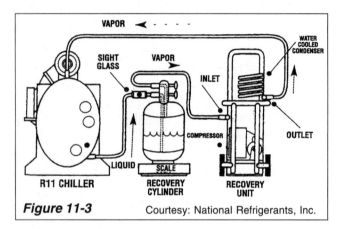

Figure 11-3   Courtesy: National Refrigerants, Inc.

The pumping rate will increase if the chiller pressure is raised to 5 to 10 psig by raising the water temperature. The rate will also improve if a ∫" or 1" connection is used at the chiller.

Vapor Recovery

Once all the liquid has been recovered, the remaining vapor can be extracted with the recovery unit as shown in Figure 11-4.

The recovery unit utilizes a water cooled condenser which requires approximately 3 gpm of water between 40°F and 75°F.

As with liquid recovery, the recovery time is improved if the connections at the chiller are changed to ∫"or 1".

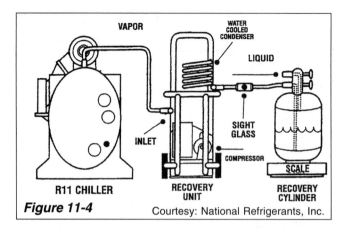

Figure 11-4   Courtesy: National Refrigerants, Inc.

R-11 Vapor must be recovered to meet EPA requirements

The liquid recovery process alone on low pressure chillers does not satisfy the demands of the law. An average 350 ton CFC-11 chiller with the liquid refrigerant recovered will still contain over 100 pounds of R-11 vapor at atmospheric pressure. A typical R-11 chiller has 3 pounds of refrigerant per ton of refrigeration. Usually after liquid recovery, 10% of the R-11 remains as vapor at atmospheric pressure.

Assuming a recovery unit has a capacity of 29" Hg vacuum, the remaining total charge would be less than 1% which would meet EPA requirements as shown in Figure 11-5.

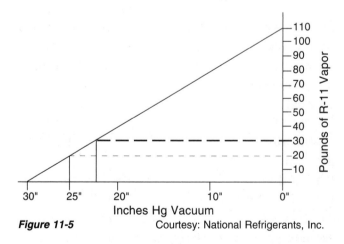

Figure 11-5   Courtesy: National Refrigerants, Inc.

Self-contained recovery systems are available with 3,400 to 5,000 pound liquid storage capacity. Liquid level indicators are available for all sizes, with some manufacturers offering electrical heaters to recycle and force liquid from the pump out tank to chiller.

All directions and guidelines for liquid and vapor recovery methods are to be taken in prescribed stages per directions in the equipment manual. The importance of following safety guidelines and practices cannot be overstressed.

Charging

Refrigerant charging is done in the evaporator section. During system charging, refrigerant gas is initially charged into the low pressure chiller before introducing liquid into its chiller barrel to prevent liquid vaporization, thereby preventing water freeze-up within the tubes. Do not inject liquid refrigerant during charging until the saturation temperature is above 32°F.

Chiller at Idle

When low pressure chillers are left idle, there must be a slight positive pressure left on the system. Heating blankets, resistance heaters, or warm water circulation through the chiller are all methods to maintain positive pressure.

Safety

ASHRAE Standard 15 requires the use of room sensors and alarms to detect refrigerant leaks. This standard includes all refrigerant safety groups. Each machine room shall activate an alarm and mechanical ventilation before the refrigerant concentrations exceed the Threshold Limit Values (TLV) and Time Weighted Averages (TWA).

Every refrigeration system shall be protected with a safety relief device or some other means designed to safely relieve pressure. Pressure relief valves are always installed in parallel, never in series. Always vent to outdoors.

Replacement Refrigerants (See Table 1)

In general, an alternative refrigerant cannot replace an existing refrigerant through any "drop-in" procedure. Consult the equipment manufacturer or the owner for maintenance records, information about changes in system design, and specific procedures to be followed when retrofitting with a different refrigerant. Perform base-line testing of the unit, determining its performance prior to retrofit.

Table 1

Acceptable Subsititues for Air Conditioning under the Significant New Alternatives Policy (SNAP) Program as of June 8, 1999				
Substitutes (Name used in the Federal Register)	Trade Name	CFC-11 Centrifugal Chillers	CFC-12, CFC-114, R-500 Centrifugal Chillers	CFC-12, R-500, Reciprocating Chillers
HCFC-123	123	R, N	N	
HCFC-22	22	N	N	N
HCFC-124	124		R, N (CFC-114 only)	
HFC-134a	134a	N	R, N	R, N
HFC-227ea		N	N	N
HFC-236fa			R, N (CFC-114 only)	
R-401A	MP-39			R, N
R-401B	MP-66			R, N
R-406A	GHG		R, N (R-500 only)	
R-409A (HCFC Blend Gamma)	409A			R, N
4-411A, R-411B	411A, 411B			R, N
FRIGC (HCFC Blend Beta)	FRIGC FR-12		R, N (CFC-12, R-500 only)	R, N
Free Zone (HCFC Blend Delta)	Freezone/R B-276		R, N (CFC-12, R-500 only)	R, N
Blend Zeta	Ikon			
Hot Shot	Hot Shot, Kar Kool		R, N (CFC-12, R-500 only)	R, N
GHG-X4	GHG-14, Autofrost Chill-it		R, N (CFC-12, R-500 only)	R, N
GHG-X5	GHG-X5		R, N (CFC-12, R-500 only)	R, N
GHG-HP (HCFC Blend Lambda)	GHG-HP			R, N
Freeze 12	Freeze 12		R, N (CFC-12, R-500 only)	R, N
G2018C	411C		R, N (CFC-12, R-500 only)	R, N
HCFC-22/HCFC-142b				(CFC-12 only)
Ammonia Vapor Compression		N	N	
Evaporative Cooling		N	N	N
Desiccant Cooling		N	N	N
Ammonia/Water Absorption		N	N	
Water/Lithium Bromide Absorption		N	N	

Key: R=Retrofit Uses. N=New Uses

Compatibility

Some alternative refrigerants are not compatible with original system components or their replacement parts. Refer to manufacturers' records to determine materials of construction. Review system maintenance records for material involved with rebuild and repair. Modification made to original system components should be recorded. Confirm with the original manufacturer that materials of construction are compatible with the new refrigerant. Refrigerant loss and equipment damage could result if these checks are not made.

Safety and Handling Procedures
PAFT

Through fifteen CFC producers, the Program for Alternative Fluorocarbon Toxicity Testing (PAFT) was sponsored. This cooperative research effort integrates past and present toxicological information to perform a careful risk assessment of proposed HCFC and HFC refrigerant alternatives to CFCs.

Testing has shown that HCFCs and HFCs exhibit properties and performance characteristics similar to CFCs, but with greatly reduced environmental impact. Acceptable levels of toxicity, in-use stability, non-flammability, and low photo-chemical reactivity are offered with the new replacements. They are not "drop-in" replacements, but they require only minimal changes to equipment when compared to other products which are not of the same "kind."

Zeotropic mixtures R-402A, R-402B, R-401A, and R-401B are classified by ASHRAE Standard 34-1992 Safety Group Clasification rating A-1.

R-123

PAFT-1 results indicate that there is nothing to prevent the use of HCFC-123 as a replacement for CFC-11 for use in general industrial areas. **The ASHRAE safety rating for R-123 is B-1.** This signifies R-123 has evidence of toxicity at concentrations below 400 ppm. Extra precautions and machine room requirements must be observed.

Although R-123 is compatible with most construction materials used with R-11, it is a somewhat stronger solvent. Elastomers made of Buna N or neoprenes, as well as some motor insulation materials and other components in hermetic motors, can become damaged. Retrofitting existing R-11 chillers to R-123 may require the replacement of gaskets, seals, and other system components to prevent leaks and to aquire proper system operation.

R-134a

PAFT-2 Program on HFC-134a has been completed. All available results show that HFC-134a is at least as safe to use as CFC-12. Ozone-depletion potential of HFC-134a is zero because it has no chlorine. This removes R-134a from the ozone-depletion controversy. Primarily due to lubrication requirement, it is not a drop-in replacement for R-12.

R-124

The PAFT-3 program on HCFC-124 began late in 1989. No adverse toxicological findings have been reported. According to existing test results, R-124 is an environmentally acceptable and functionally equivalent replacement for R-114. This refrigerant carries ASHRAE Standard 34 Safety Group Classification rating of A-1.

Common Safety Related Questions
Flammability
Are HCFCs and HFCs flammable?
Some HCFCs and HFCs are flammable. For example, HFC-152a and HFC 32 are flammable refrigerants if used alone. However, both refrigerants when used in a blend, are not flammable even when fractionation occurs.

HFC-134a has demonstrated combustibility at test pressures as low as 5.5 PSIG (at 350°F) when mixed with air at concentrations of generally more than 60 percent (volume) air. Higher pressures are required at lower temperatures for combustibility.

This refrigerant should not be mixed with air for leak testing. It should not be allowed to be present with high concentrations of air above atmospheric pressure.

Inhalation
Are HCFCs and HFCs toxic?
They pose no hazard of systemic toxicity, carcinogenicity, mutagenicity, or teratogenicity at or below Time Weighted Average (TWA) of air borne concentration exposures for full work shifts (eight or 12 hour TWA).

Short term exposures should not exceed three times the assigned exposure limit for more than a total of thirty minutes during a work day, provided that the TWA is not exceeded. **Under no circumstances should they exceed five times the TWA.**

Property Comparisons

	CFC-11	HCFC-123	CFC-12	HFC-134a
Molecular structure	CCl_3F	$CHCl_2\text{-}Cf_3$	CCl_2F_2	$CH_2F\text{-}Cf_3$
Boiling Point °C (°F)	24 (74.9)	28 (81.7)	-30 (-21.6)	-26 (-15.7)
Toxicity, ppm (v/v)	1000 (TLV) ✗	10 (AEL)≠	1000 (AEL)	1000(AEL)
COP	7.57 ◑	7.34 ◑	2.27 ↘	2.22 ↘
Ozone depletion potential	1.0	0.0015	1.0	0.0
Global warming potential	1.0	0.002	3.0	0.28

✗ A Threshold Limit Value (TLV), established for industrial chemicals by the American Conference of Governmental Industrial Hygienists, is the time-weighted average concentration of an airborne chemical to which nearly all workers may be exposed during an 8-hour day, 40-hour week without adverse effect.

≠ An acceptable Exposure Limit (AEL) is the recommended time-weighted average concentration of an airborne chemical to which nearly all workers may be exposed during an 8-hour day, 40-hour week without adverse effect, as determined by the DuPont Company for compounds that do not have a TLV.

◑ 40ϓF/100ϓF/0ϓF SH

↘ -10ϓF/130ϓF/100ϓF SH (Compressor suction temperature at 90ϓ F.)

What are the symptoms of exposure above the AEL?

Inhalation initially attacks the central nervous system, creating a narcotic effect. A feeling of intoxication and dizziness with loss of coordination and slurred speech are symptoms. Cardiac irregularities, unconsciousness, and ultimate death can result from continued breathing of this concentration.

If any of these symptoms become evident, move to fresh air and seek medical help immediately. Recovery can occur quickly, with little aftereffect (see Property Comparisons chart on page 135 showing chemical formulas, boiling points, and AELs).

What is cardiac sensitization?

The heart can become sensitized to adrenaline if vapors in excess of the AEL are inhaled. Cardiac irregularities and cardiac arrest could result. Likelihood of this occurring increases if you are under physical or emotional stress.

What can I do if I am near someone experiencing this reaction?

Medical attention must be given immediately. Do <u>not</u> treat with adrenaline (epinephrine) or similar heart stimulants. An increased risk of cardiac arrhythmia and cardiac arrest will result. Administer oxygen if the person is having difficulty breathing. Administer artificial respiration if breathing has stopped, preferably mouth-to-mouth. If a large release of vapor occurs, such as during a large spill or leak, vapors may concentrate and displace oxygen available for breathing. This would cause suffocation.

What are Emergency Exposure Limits?

Emergency Exposure Limits (EELs) are set for emergency situations, such as a spill or accidental release of a chemical. EELs specify brief durations and concentrations from which escape is feasible without any irreversible effects on health. EELs are only applicable to emergency situations where reoccurrence is expected to be rare in an individual's lifetime. Refer to chart on page 136.

How do refrigerant vapors concentrate and become dangerous?

CFC, HCFC, and HFC refrigerants are three to five times heavier than the air we breath. Leaking vapors tend to collect and concentrate near the floor or in low depressions, displacing air. High concentrations of refrigerant means less air.

How can I work safely on systems in enclosed areas?

❑ Purge and vent relief piping must exit out of doors.

❑ Equipment area must be well ventilated to remove any refrigerant vapors, preferably at low level.

❑ Install air-monitoring equipment to detect leaks. Refer to ASHRAE Standard 15 for ventilation requirements.

Is deliberate inhalation of refrigerant vapors dangerous?

This practice is extremely dangerous, as death can occur.

Can I smell these refrigerants?

Some have a slightly sweet odor, such as R-123, that cannot be detected at levels considered safe for exposure. Frequent leak checks and installation of permanent detectors are often necessary for enclosed areas used by personnel.

Are these refrigerants toxic when taken orally?

Ingestion should be avoided. If accidental ingestion occurs, do not induce vomiting. Induced vomiting is dangerous because when vomiting occurs, refrigerant may draw into the lungs leading to other very serious physical hazards. Get medical attention immediately.

Is skin or eye contact with these refrigerants hazardous?

When these refrigerants are in liquid form, serious damage to skin and eyes can occur. If the liquid experiences direct expansion, frostbite can occur. Refrigerants which are liquid at room temperature tend to dissolve the skin's protective fat, causing dryness and irritation, particularly after prolonged contact.

How should I braze or weld pipes on air conditioning or refrigeration systems?

❑ Assure adequate ventilation in the area. Provide auxiliary ventilation if necessary.

❑ Recover refrigerants out of the system into cylinders. Make sure refrigerant is not vented into work space.

❑ Purge system with nitrogen.

Allowable Personal Exposure Levels for HCFC-123 and CFC-11			
Exposure Level	**Source**	**HCFC-123**	**CFC-11**
Permissible Exposure Limit (PEL)	OSHA	Not Established	Not Established
Ceiling	OSHA	Not Established	1,000 ppm
Acceptable Exposure Level (AEL)	DuPont	30 ppm	Not Established
Emergency Exposure Level (EEL)	DuPont	1,000 ppm	Not Established

Permissible Exposure Limit. OSHA time-weighted average limit. The employee's average airborne exposure in any 8-hour workshift of a 40-hour workweek, which shall not be exceeded.

Ceiling. The employee's exposure level, which shall not be exceeded for any time period.

Acceptable Exposure Level. The employee's average airborne exposure in any 8-hour workshift of a 40-hour workweek, which shall not be exceeded.

Emergency Exposure Limits. The concentration from which escape is feasible without any irreversible effects on health in an emergency situation where reoccurrence is expected to be rare in an individual's lifetime.

❑ Leave system open before beginning welding or brazing operations. This will prevent possibility of hydrostatic pressure within the system, and potential refrigerant decomposition.

Refrigerant Decomposition

When refrigerants are exposed to high temperatures from open flame or resistive heater elements, decomposition occurs. Decomposition produces toxic and irritating compounds, such as hydrogen chloride and hydrogen fluoride.

Are products of decomposition hazardous?

Yes. Acidic vapors produced are dangerous and the area should be evacuated and ventilated to prevent exposure to personnel. Anyone exposed to products of decomposition should be taken to fresh air and given medical treatment immediately. The irritating nature of the fumes will generally force people to leave the area well in advance of hazardous conditions.

Equipment Room/Job Site Requirements

1. Maintain equipment rooms in compliance with local safety codes applicable to your area. Refer to ASHRAE Standard 15R where local codes are not yet outlined to accommodate alternative refrigerants. ASHRAE standard 15R further defines five additional areas that should be conformed to for all refrigerants:

❑ **MONITORS** Each machinery room shall contain a dectector, located in an area where refrigerant from a leak will concentrate, which shall activate an alarm and cause mechanical ventilation to operate in accordance with 8.13.4 at a value not greater than the corresponding TLV-TWA (or toxicity measure consistent therewith).

Exception: For ammonia refers to 8.14(g).

❑ **ALARMS** An alarm which activates at, or below, the AEL for Group B1 refrigerants. When used, an oxygen alarm will be activated at not less than 19.5 percent by volume.

❑ **VENTILATION** Mechanical ventilation must be sized and used per ASHRAE Standard 15R. This is typically not required for penthouse and lean-to-applications.

❑ **PURGE VENTING** Rupture disks and purges must be vented out of doors, using refrigerant compatible materials. A drip-leg and shutoff valve should be provided on vent piping.

❑ **BREATHING APPARATUS** At least one approved self-contained breathing apparatus for emergency use should be at a convenient location within or very near an equipment room.

2. Periodically check oxygen/refrigerant monitor for change of drift of levels in the equipment room. In the absence of monitoring equipment, or if high concentrations are indicated (exceeds applicable exposure limits), *VENTILATE* the equipment room to ensure adequate supply of fresh air.

3. Review job site records:
   a. For recent monitoring levels.

   b. For integrity of monitoring equipment and calibration intervals versus manufacturer recommendations (generally monthly).

   c. For operational history and service record of the chiller.

   d. For availability of self-contained breathing apparatus and other personal safety equipment. Devices must be calibrated, tested, and/or maintained in proper state.

Safety Group Classification System

In accordance with toxicity and flammability potential, refrigerants are coded within ASHRAE Standard 34 as follows:

Toxicity is indicated with a capital letter indicating level. From the base of "A" to "B", the letter indicates greater toxicity.

Class A signifies refrigerants for which toxicity has not been identified at concentrations less than or equal to 400 ppm, based on data used to determine Threshold Limit Value-Time-Weighted Average TLV-TWA or consistent indices.

Class B signifies refrigerants for which there is evidence of toxicity at concentrations below 400 ppm, based on data used to determine TLV-TWA or consistent indices.

Incrementally, the value of one through three is employed to indicate flammability. Tests are made in accordance with ASTM E681-85 except that ignition source is an electrically-activated kitchen match head for hydrocarbon refrigerants.

Class 1 indicates refrigerants that do not show flame propagation when tested in air at 14.7 psia and 65°F.

Class 2 signifies refrigerants having a lower flammability limit (LFL) of more than 0.00625 lb/ft³ at 70°F and 14.7 psia, and a heat of combustion of less than 8,174 Btu/lb.

Class 3 indicates refrigerants that are highly flammable, as defined by an LFL of less than or equal to 0.00625 lb/ft³ at 70°F and 14.7 psia, or a heat combustion greater than or equal to 8,174 Btu/lb. See Table 2 and Table 3.

Retrofitting of Chillers with Alternative Refrigerants

Many of the existing centrifugal chillers are charged with low pressure R-11. Thousands of chillers with an average service life of 30 years are now operating in America. In order to use its replacement, equipment manufacturers require close communication with the manufacturer of the given assembly. The first step toward success in refrigerant change-over and an efficient retrofit should be to contact the original equipment manufacture (OEM) to ensure that proper procedures are followed and compatible materials are used.

Table 2
Safety Matrix

	Low Toxicity TLV => 400 ppm	Higher Toxicity TLV =< 399 ppm
High Flame	A3 R-290 (Propane)	B3 R-1140
Mild Flame	A2 R-152a	B2 R-717 (Ammonia)
No Flame	A1 R-11 R-12 R-22 R-134a R-404A R-410A R-401A/B R-402A/B	B1 R-123

Note that, of the refrigerants now commonly in use or coming into use as replacements, most are classified as A1, but R-123 is B1.

For additional information regarding Safety Codes for Mechanical Refrigeration, use ANSI/ASHRAE Standard 15.

The use of low pressure refrigerant R-123 in place of R-11 in retrofit may involve replacement of seals, gaskets, bushings, diaphragms, and motor insulation (or the motor field assembly). Comprehensive engineering may also be necessary for efficiency improvement of heat exchange surfaces and compressor operation. Controls may also have to be modified and improved.

Table 3

Safety groups for Applicable Low Pressure Refrigerants *ASHRAE Standard 15-1992*		
Refrigerant	**Name**	**Safety Group**
CFC 11	Trichlorofluormethane	A1
CFC 113	1,1,2-trichlorotrifluoroethane	A1
CFC 114	1,2-dichloretraflouroethane	A1
HCFC 123	2,3-dichloro-1,1,1-trifluoroethane	B1

Presently, only R-123 can be used in existing equipment designed for R-11. Normal system operating ranges of saturated pressures compare very closely when saturated pressure versus temperature curves are referred to.

Relief from Retrofit/Retirement

The owners or operators of industrial process refrigeration equipment or federally owned chillers may be relieved from the retrofit or repair requirements if:

❑ second efforts to repair the same leaks that were subject to the first repair efforts are successful, or

❑ within 180 days of the failed follow-up verification test, the owners or operators determine the leak rate is below 35 percent. In this case, the owners or operators must notify the EPA as to how this determination will be made, and must submit the information within 30 days of the failed verification test.

R-123 Performance

Considering identical chilling applications, the new refrigerant has a higher specific volume and lower acoustic velocity than R-11. Regarding these two factors, the difference in this comparison is significant.

Given the same capacity requirements, an R-123 compressor will have to circulate 10-15 percent more inlet cubic feet per minute (ICFM) and generate about seven percent more lift, or head. Net effect of these two changes could be an average 10-15 percent reduction in capacity on the compressor. Actual test results with various compressors have shown losses ranging from 0-18 percent.

When equipment designated for R-123 is assembled as new, the compressor is selected to match requirements of R-123 resulting in a larger compressor than for R-11. For retrofit of older equipment, it is important that the OEM is consulted to determine potential capacity losses and select an appropriate package for the conversion. By working with equipment manufacturers and following sound engineering practices, equipment can be converted to employ the new refrigerant with minimal losses in capacity and performance.

R-123 Compatibility and Handling

R-123 has stronger solvent properties than R-11 and may damage elastomers made of Buna N (or commonly used neoprenes), as well as motor insulations and other components in hermetic motors. Open motor designs do not have this problem. Original equipment manufacturers have identified and tested acceptable materials to replace seals, bushings, gaskets, and other components in retrofit of older equipment.

Refrigerant purge losses on large centrifugal equipment can be significant. Installation of high efficiency purge units is very important to prevent evacuation of large amounts of refrigerant during operation. These high efficiency purge systems can reduce refrigerant purge losses by 97 percent or more.

Little difference in handling requirements exist between R-123 and R-11. Currently, there is an assigned allowable exposure limit (AEL) for R-123 of 30 parts per million (ppm), which is below that normally associated with CFCs. Requirements for alarm systems and ventilation are outlined in ASHRAE Standard 15. Relief valve discharge headers and purge devices should be routed to discharge out of doors and away from all air intakes.

Retrofit of R-134a into Existing Equipment

Performance of medium pressure refrigerant R-134a is similar to R-12 for standard comfort cooling application. Minimal sacrifice in operational efficiency or capacity is possible. For larger equipment operated through a speed reduction box, drive gear changes are likely.

As requirements demand lower evaporating temperatures and increased condensing temperatures, there is a separation of pressures which could significantly affect performance. To attain desired performance at low temperature applications, modification may be necessary to increase compressor speed or achieve greater lift characteristics.

Standard chiller application in the range of chilled water has proved an increase in Net Refrigerating Effect (NRE), with an offsetting increase in specific volume of the refrigerant entering the compressor. Variable operating conditions cause a shift in performance which can be either favorable or negative.Higher acoustic velocity and greater operating specific volume of the suction gas results in a significantly higher required compressor lift, or head. Increases in required speed of an R-134a chiller can equate to four to eight percent (in comparison to a system designed for R-12).

Oil/Refrigerant Relationships

Current oils used with R-11 and R-12 are fully soluble over the range of expected operation conditions. Refrigeration systems employing R-11 or R-12 take advantage of this full solubility when considering oil return.

Oil and refrigerant solubility is limited, with R-134a in particular presenting an "inverse" solubility curve. Instead of becoming insoluble at low temperatures, these solutions tend to separate at HIGH temperatures. Instead of having a tendency to congeal in the evaporator, the oil is more likely to separate out in the condenser or liquid line receiver. Due to poor solubility of mineral-based oils with R-134a and proven inferior lubrication qualities resulting from this combination, any retrofit of R-134a from R-12 must include a switch to a synthetic Polyol Ester refrigeration lubricant.

R-12/R-134a Procedure for Retrofit; Polyol Ester Refrigerant Oil

❑ Change original compressor lubricant to Polyol Ester (POE).

❑ Operate system with R-12 for 24 hours to circulate the POE throughout the system and flush the mineral based system oil back to the compressor.

❑ Repeat the above procedures until residual oil level is below 5%.

❑ Remove R-12 from the complete system.

❑ Replace filter/drier and perform any manufacture recommended changes to the system.

❑ Evacuate the system to manufacturer's specifications.

❑ Recharge system with the proper amount of R-134a (usually 85% - 90% of the original R-12 charge).

❑ Compare the operating system to the original operational data. Adjust controls and/or refrigerant system charge for correct operation.

❑ Label system with the new refrigerant and oil type.

Note: Replacement of thermostatic expansion valve, filter dryer, and alteration of pressure control settings to appropriate values are necessary preliminary steps to refrigerant retrofit.

Planning and Acting for the Future

Observing the refrigerant recycling regulations for Section 608 is essential in order to conserve existing stocks of refrigerants, as well as to comply with Clean Air Act requirements. However, owners of equipment that use CFC refrigerants should look beyond the immediate need to maintain existing equipment. Owners are advised to begin the process of converting or replacing existing equipment with equipment that uses alternative refrigerants.

To assist owners, suppliers, technicians, and others involved in comfort chiller and commercial refrigeration management, the EPA has published a series of short factsheets and expects to produce more material. Copies of material produced by the EPA Stratospheric Protection Division are available.

For further information concerning regulations related to stratospheric ozone protection, please call the EPA Stratospheric Ozone Protection Hotline at 800 296-1996. The Hotline is open between 10:00 A.M. and 4:00 P.M., Eastern Standard Time.

Notes

Notes

History of Refrigerants

Even though water and ice were the first refrigerants, ether was the first commercial refrigerant. In 1850, ice was made by evaporating ether under a vacuum produced by a steam driven pump. By 1855, there was an ether machine that could produce a maximum of 2,000 pounds of ice per day. This was a vapor compression process that used a volatile etheral fluid as the refrigerant. The refrigerant was then condensed and reused generating no wasted ether. Many other etheral based machines were developed including one that transported chilled meats across the sea from France to South America by ship. Because ether operated in a vacuum and was extremely flammable, the last ether machine was made in 1902.

The following decades saw numerous compounds tested as refrigerants. In fact, one can safely say that just about every volatile fluid was tested as a refrigerant. Mechanical vapor compression refrigeration was firmly established by the turn of the century. However, these early refrigerants all had disadvantages and advantages. Listed below are some of the most popular refrigerants used according to date.

1850-1856

An ammonia (R-717) machine received a patent. Good thermodynamic properties and low costs made ammonia a widely used refrigerant even today. Ammonia is very irritable to the mucous membranes and is flammable in certain concentrations.

1882-1886

Carbon dioxide (R-744) was used on British ships into the 1940s until it was replaced by chlorofluorocarbons. Carbon dioxide systems never saw widespread usage in the United States.

1880-1940

Sulfur dioxide (R-764) has an advantage of low costs and low operating pressures for warmer climates, but high enough pressures to remain out of a vacuum. Sulfur dioxide was used as a refrigerant in household refrigerators around 1900. One drawback of sulfur dioxide systems was that the refrigerant reacts with moisture and forms sulfurous acid. This resulted in a lot of seized compressors. Even though sulfur dioxide is a toxic refrigerant, the smallest of leaks can be detected by smell. This refrigerant remained popular in smaller units until the 1940s.

1890

Methyl chloride (R-40) was used sparingly in the United States until 1910. Its earlier use was in shipping meats across the sea. Methyl chloride has a sweet, etheral odor and has somewhat of an anesthetic effect when inhaled. It is also mildly flammable. Leaks in large systems had many fatal results. This refrigerant also reacted with the aluminum in hermetic motors in the 1940s. Its use declined in the late 1930s.

1880-1890

Ethyl chloride (R-160) was used as an anesthetic. Liquid refrigerant was sprayed on the skin before surgery. This refrigerant was used in some household refrigerators after 1900.

Some other refrigerants that were experimented with and often used for short periods of time included:

methylamine	**naphtha**
nitrous oxide	**methyl acetate**
butane	**pentane**
propylene	**isobutylene**
carbon tetrachloride	**gasoline**
dielene	**trielene**
ethyl bromide	

Most of the refrigerants mentioned so far are either toxic, flammable, or smelled horrible. There was always some health risk when incorporating these refrigerants in the home. The refrigeration industry needed newer, safer refrigerants. The Frigidaire Company asked General Motors Research Laboratories to develop a safe refrigerant. This was the beginning of the chlorofluorocarbon (CFC) refrigerants. The research team, headed by Thomas Midgley of General Motors, settled on R-12 as the refrigerant most suitable for commercial use. These refrigerants were all either methane or ethane based, and are explained in Section Two. The first use of R-12 was in small ice cream applications in 1931. R-12 soon became a commercial refrigerant for room coolers. In 1933, R-11 was used often in centrifugal compressors for air conditioning applications.

The years to follow brought on the development of many more chlorofluorocarbon refrigerants. Listed below are these refrigerants with the dates of introduction to the commercial market along with other important dates.

1930	Development of chlorofluorocarbons
1931	R-12
1932	R-11
1933	R-114
1934	R-113
1936	R-22
1961	R-502 An azeotropic mixture of HCFC-22 and CFC-115.
1974	Ozone depletion theory

1978	Ban on non-essential aerosols.
	Global warming came into view.
1985	Stratospheric ozone hole discovered.
1987	Montreal Protoco
	Current tax rate schedule on CFC refrigerants.
1990	Clean Air Act Amendments
	Refrigerant production cuts and bans.
July 1992	Unlawful to vent CFCs and HCFCs into atmosphere.
Nov. 15, 1995	Unlawful to vent alternative refrigerants (HFCs) into the atmosphere.
1996	Phaseout of CFC refrigerants
1996	Freeze HCFC production
1997	Kyoto Protocol intended to reduce world wide global warming gasses. Global Warming has become a major environmental issue.
1998	EPA "proposed" more strict regulations on recovery/recycling standards, equipment leak rates, and alternative refrigerants.
2020	No production and no importing of HCFC-22 (R-22).
2030	No production and no importing of any HCFC refrigerant.

Before we start, let's talk about test anxiety. We all fear being questioned or being required to answer questions for fear of being wrong. No one wants to be wrong. To be wrong suggests failure, and in the case of the matter at hand, failure of the EPA approved exam means no certification, and no certification means we don't work in this industry. WOW! Scary.

Now let's be rational. If we as technicians have the fundamental knowledge and skills to be able to install, maintain, and service air conditioning and refrigeration equipment, we are not a failure. If we add to that knowledge the changes (laws) that have transpired because of stratospheric ozone depletion and global warming as found in this manual, we can be successful in "passing" not "failing" the exam. All we may lack is the confidence to test our capabilities, and if we can put these painful anxieties of testing aside momentarily while we learn the new material and refresh our memory with what we already know, we can do it. Even if we haven't taken a multiple choice exam in many years, we can pass the EPA test.

POINTERS TO HELP LESSEN YOUR TEST ANXIETIES:

- Be **well rested**. The mind of a tired body cannot function at its peak.

- Have a **positive attitude**. It is important to psych yourself up mentally that you **can** do it.

- Be in a **comfortable setting**. The room where you are taking your test should have good lighting, be free of distractions, and comfortable so that you can **relax** in "your space" for the test taking.

- **Take your time**. Ample time will be provided for the exam. Don't be forced to hurry yourself with the test. Also, don't allow yourself to become frustrated or anxious with filling out the score sheets with their narrow columns, lines and spaces. Be sure to darken the circles completely—with confidence.

- **Take deep breaths**. Long, deep breaths will be helpful to your **relaxing.**

- **Read the question**. You have to believe that it was not intended by the EPA for the question to be tricky and that it was written to check your knowledge and skills in the matter at hand. The EPA is testing your knowledge regarding stratospheric ozone depletion and global warming and your competency with refrigeration and air conditioning systems so as to assure containment, conservation, and recycling of Class I & II substances. **Do not read too much into the question.**

- **Review the four answers** (A-D) **for the best response**. If the answer is not obvious to you, select an answer through the process of elimination. One to three of the responses are distractors and can in some cases be eliminated, leading you to the correct answer.

- If you are not sure, answer the question with your **initial reaction**. Write the number of this question down on a scrap piece of paper and after you have answered all of the questions, take a deep breath, collect your thoughts and go over these questions again. Other questions you have answered may trigger different thoughts (hopefully the right ones) regarding what is being sought by the EPA in the question.

- **Historically, your first reaction is the correct answer**. We don't recommend reviewing the entire test and "trying to read the question differently" and then changing your answer.

- **RELAX.**

- **GOOD LUCK!**

Following are forty sample questions divided by types to familiarize yourself with the type of questions you will see as well as practice and test your ability.

SAMPLE TEST

Correct answers are listed on bottom of page 151

CORE QUESTIONS

1. As of what date did it become unlawful to release Class I and Class II refrigerants into the atmosphere?
   A. July 1, 1992
   B. July 1, 1993
   C. November 14, 1994
   D. January 1, 1996

2. The atom found in CFC and HCFC refrigerants that destroys ozone in the stratosphere is:
   A. Fluorine
   B. Carbon
   C. Hydrogen
   D. Chlorine

3. Which refrigerant is a CFC?
   A. R-134a
   B. R-123
   C. R-22
   D. R-12

4. Which refrigerant is a HFC?
   A. R-134a
   B. R-123
   C. R-22
   D. R-12

5. Which refrigerant contains no chlorine?
   A. R-134a
   B. R-123
   C. R-22
   D. R-12

6. The rule of thumb for refilling approved cylinders is a maximum of ____ percent liquid.
   A. 60%
   B. 70%
   C. 80%
   D. 90%

7. To RECOVER refrigerant is to:
   A. Remove refrigerant in any condition from a system in either an active or passive manner, and store it in an external container without necessarily testing or processing.
   B. Reduce contaminants in used refrigerant by oil separation through filter driers.
   C. Reprocess refrigerant to new product specifications.
   D. Remove refrigerant and change ownership.

8. R-134a is a "drop-in" refrigerant for:
   A. R-12
   B. R-22
   C. R-11
   D. R-134a is not a drop-in refrigerant

9. The condition and state of the refrigerant leaving a receiver is:
   A. subcooled liquid
   B. subcooled vapor
   C. superheated vapor
   D. superheated liquid

10. The component of an air conditioning system that changes a low pressure vapor to a high pressure vapor is the:
    A. condenser
    B. metering device
    C. evaporator
    D. compressor

Type I Questions

11. Which best describes the definition of Type I "Small appliance", as defined by the EPA?
    A. Systems manufactured, charged, and hermetically sealed with five (5) pounds or less of refrigerant
    B. Refrigerators, freezers, room air conditioners, and central air conditioners
    C. Any appliance charged with more than five (5) pounds of refrigerant
    D. Any appliance charged with less than two (2) pounds of refrigerant

12. For small appliance use, the recovery equipment manufactured after November 15, 1993 must be capable of recovering:
    A. 80% of the refrigerant when the compressor is not operating or achieve a 4 inch vacuum under ARI 740-1993
    B. 90% of the refrigerant when the compressor is operating or achieve a 4 inch vacuum under ARI 740-1993
    C. 99% of the refrigerant regardless of compressor operation and achieve a 10 inch vacuum under ARI 740-1993
    D. Both A and B

13. The sale of Class I and Class II refrigerants will be restricted to technicians certified by an EPA approved program after:
    A. July 1, 1992
    B. November 15, 1993
    C. August 12, 1993
    D. November 14, 1994

14. The release of vapor from the top of a graduated charging cylinder when filling may:
    A. Be vented to the atmosphere
    B. Be vented if the quantity does not exceed three (3) pounds
    C. Not be vented and must be recovered
    D. Be vented, but not inhaled

15. Should regulations of the Clean Air Act (CAA) change after a technician is certified:
    A. The technician must take a new test to be recertified
    B. All technicians who previously passed with an 80% will be grandfathered
    C. It will be the technician's responsibility to learn and comply with future changes in the law
    D. The technician must be retested and pass the exam with an 84%

16. System dependent (passive) refrigerant recovery of small appliances:
    A. Do not require an operating compressor
    B. Requires 80% of the refrigerant to be recovered
    C. Recovers refrigerant in a non-pressurized container
    D. All of the above

17. Before disposing of a small appliance containing R-12, it is necessary to:
    A. Pressurize with nitrogen
    B. Recover the refrigerant
    C. Turn upside down
    D. Thoroughly leak check

18. A system has been operating with a complete restriction at the capillary tube inlet, what access is required for recovery?
    A. One access valve on the low side of the system
    B. Two access valves, high and low side of system
    C. One access valve on the high side of the system
    D. One access valve on the evaporator, and one on the low side

19. CFCs will not be manufactured in the United States after:
    A. 2000
    B. 1995
    C. 2005
    D. 1996

20. To work on small appliances after November 14, 1993, a technician must be certified as:
    A. Type I
    B. Type II
    C. Universal
    D. Either A or C

Type II Questions

21. Which refrigerant can be used for leak detection as a trace gas and pressurized with nitrogen?
    A. R-12
    B. R-11
    C. R-22
    D. R-115

22. Traces of oil around the sight glass inlet fitting of a refrigeration system might be the indication of:
    A. A leak
    B. Excessive oil in the system
    C. An overcharge
    D. A restriction at the TXV

23. Type II classification, as identified by the EPA, applies to what equipment?
    A. Small appliances with five (5) pounds of refrigerant or less
    B. Refrigerants, freezer, and vending machines appliances
    C. Low pressure appliances
    D. Split air conditioning equipment with five (5) pounds of refrigerant or greater

24. The required level of evacuation for recovery equipment manufactured after November 15, 1993, on a system containing less than 200 pounds of R-12 refrigerant is:
    A. 0 inches Hg
    B. 4 inches Hg
    C. 10 inches Hg
    D. 15 inches Hg

25. Industrial process and commercial refrigeration equipment with over 50 lbs. of refrigerant with an annual leak rate of ____% requires repair under EPA regulations.
    A. 0
    B. 15
    C. 35
    D. 50

26. Comfort cooling chillers and all other equipment with over 50 lbs. of refrigerant with an annual leak rate of ____% requires repair under EPA regulations.
    A. 0
    B. 15
    C. 35
    D. 50

27. The majority of the liquid to be recovered from a system will be found in the:
    A. Condenser
    B. Receiver (when applied)
    C. Low side of system
    D. Evaporator

28. It becomes the owner's responsibility to maintain records of all refrigerant added to units that contain more than ____ pounds of refrigerant charge.
    A. 15
    B. 20
    C. 35
    D. 50

29. Exceptions to the required evacuation levels for recovery equipment that require an appliance be evacuated to only 0 psig apply to appliances that:
    A. Are being salvaged
    B. Are filled with water or substances that would damage the recovery equipment
    C. Have defective evaporator fan motors
    D. Have air cooled condensers

30. The condition and state of refrigerant entering the receiver is:
    A. Superheated high pressure vapor
    B. Superheated low pressure vapor
    C. Subcooled high pressure liquid ✓
    D. Subcooled low pressure liquid

Type III Questions

31. Recovery machines using water as the condensing medium would generally use the:
    A. Cooling tower water
    B. Municipal water supply ✓
    C. Condensate water
    D. Ice water

32. Frost would be best removed from a sight glass by:
    A. Reversing the cycle
    B. Chip the ice off
    C. Spraying with alcohol ✓
    D. Turn the water supply off

33. The maximum pressure that should be applied to a low pressure chiller when leak checking with controlled nitrogen is:
    A. 3 psig
    B. 10 psig ✓
    C. 20 psig
    D. 30 psig

34. Water tube leaks in a low pressure chiller are usually found with:
    A. Water puddles
    B. Frosted coils
    C. A hydrostatic tube test ✓
    D. A leak detector

35. To prevent freezing of the water coils of a low pressure chiller, it is recommended that:
    A. When charging, begin with vapor phase
    B. Circulate water through the chiller
    C. Do not inject liquid during charging until saturation temperature is above 32 degrees
    D. All of the above

36. Under ASHRAE Standard 15, what refrigerant requires equipment room sensors?
    A. R-12
    B. R-500
    C. R-123 ✓
    D. R-134a

37. Under ASHRAE Standard 15, the following refrigerants require equipment room oxygen deprivation sensors:
    A. R-11
    B. R-12
    C. R-134a
    D. All of the above ✓

38. Low pressure chillers require purge units because:
    A. They operate below atmospheric pressure
    B. They draw non-condensables through gaskets and seals
    C. Both A and B ✓
    D. They don't require purge units

39. The purge unit draws from the:
    A. Top of the condenser
    B. Suction of the compressor
    C. Evaporator
    D. Rupture disk

40. If excessive nitrogen pressure is exerted within a low pressure chiller, what component would fail first?
    A. Evaporator coil
    B. Rupture disk ✓
    C. Compressor seals
    D. Cooling tower

Test Answers

1. A	11. A	21. C	31. B
2. D	12. D	22. A	32. C
3. D	13. D	23. D	33. B
4. A	14. C	24. C	34. C
5. A	15. C	25. C	35. D
6. C	16. D	26. B	36. C
7. A	17. B	27. B	37. D
8. D	18. C	28. D	38. C
9. A	19. B	29. B	39. A
10. D	20. D	30. C	40. B

APPENDIX III
Related Forms, Logs, and Reports

- **Virgin Refrigerant Use Log**
- **Refrigerant Removal Incident Report**
- **Accidental or Unintentional Venting Report**
- **Acquisition Certification Form with Instructions (EPA)**
  — **EPA Regional Offices**

Virgin Refrigerant Use Log

Employee Name _____   Employee No. _____   Truck No. _____

Week End Date _____   Page _____ of _____

Date	Customer Name	Job No.	Refrigerant R-12	Refrigerant R-22	Refrigerant R-500	Refrigerant R-502

	Refrigerant Types					
	R-12	R-22			Received by	Date
Empty disposable tanks turned in						
Reclaimed refrigerant returned Quantity by pound						
					Customer Name	
Virgin refrigerant given Quantity						
Date						
Quantity						
Date						

Contaminated/mixed refrigerant returned in red-tagged drum   Lbs. of Refrigerant [   ]   Drum Serial No. [   ]

Received by [   ]   Date [   ]

You will not receive virgin refrigerant without first turning in an empty drum. In addition to this form, a material requisition is still required to be completed for refrigerant given out. No refrigerant pickups are to be made from a supplier.

Distribute as follows: 1 - White - Office 2 - Yellow - Parts Dept.   3 - Pink - Technician

Refrigerant Removal Incident

Customer Name _____ Date _____

Address _____ City _____ Job No. _____

Equipment Manufacturer	Model Number	Serial Number	Refrigerant Type

Reason for Incident _____

Integrity of Refrigerant   Acid ☐ Yes ☐ No   Moisture ☐ Yes ☐ No

Refrigerant Recovery Model No. [_____]   Company Serial No. [_____]

Hour Meter Reading: Start _____ Finish _____   Company Serial No. of Recovery Drum _____

Weight of Recovery Drum: Start _____ Finish _____   Total Lbs. Refrigerant Recovered _____

Refrigerant Filtered and Dried During Removal? ☐ Yes ☐ No   Compressor Meg Reading Before Removal _____

Disposition of Refrigerant: ☐ Recycle   ☐ Reclaim   ☑ Disposal

Comments: _____

Leak Repair Report

Method Used for Leak Detection _____
_____

Materials Used for Leak Detection _____
_____

All Refrigerant Removed Prior to Putting in Nitrogen?   ☐ Yes ☐ No

Nitrogen Vented to Atmosphere?   ☐ Yes ☐ No   Customer Advised?   ☐ Yes ☐ No

Reason Leak Happened? _____
_____

Method Used to Repair Leak and Improvements Made to Keep this from Happening Again _____
_____

Technician Name _____   Employee Number _____

Customer Signature _____   Date _____

Distribute as follows: 1 - White - Office 2 - Yellow - Parts Dept.   3 - Pink - Technician

Accidental or Unintentional Venting Report

Date _____

Customer Name _____

Address _____

City _____ State _____

Job # _____

Type of Refrigerant Vented _____ Approx. How Many Pounds Were Vented _____

Description of What Happened _____

_____

_____

_____

_____

Why Did It Happen? _____

_____

_____

_____

What Precautions Have You Taken
To Prevent This From Happening Again? _____

_____

_____

Was Anyone Else Aware Of This Situation?   ☐ Yes   ☐ No

If So, Whom _____

Did You Inform The Customer?   ☐ Yes   ☐ No

Technician Name_____ Employee Number_____

Customer Signature _____ Date _____

THE UNITED STATES ENVIRONMENTAL PROTECTION AGENCY (EPA)
REFRIGERATION RECOVERY OR RECYCLING DEVICE
ACQUISITION CERTIFICATION FORM

EPA regulations require establishments that service or dispose of refrigeration or air conditioning equipment to certify [by 90 days after publication of the final rule] that they have acquired recovery or recycling devices that meet EPA standards for such devices. To certify that you have acquired equipment, please complete this form according to the instructions and **mail it to the appropriate EPA Regional Office. BOTH THE INSTRUCTIONS AND MAILING ADDRESSES CAN BE FOUND ON THE REVERSE SIDE OF THIS FORM.**

PART 1: ESTABLISHMENT INFORMATION

Name of Establishment

Street

(Area Code)Telephone Number

City   State   Zip Code

Number of Service Vehicles Based at Establishment

PART 2: REGULATORY CLASSIFICATION

Identify the type of work performed by the establishment. **Check all boxes that apply.**

☐ Type A -Service small appliances
☐ Type B -Service refrigeration or air conditioning equipment other than small appliances
☐ Type C -Dispose of small appliances
☐ Type D -Dispose of refrigeration or air conditioning equipment other than small appliances

PART 3: DEVICE IDENTIFICATION

Name of Device(s) Manufacturer	Model Number	Year	Serial Number (if any)	Check Box if Self-Contained
1.				☐
2				☐
3.				☐
4.				☐
5.				☐
6.				☐
7.				☐

PART 4: CERTIFICATION SIGNATURE

I certify that the establishment in Part 1 has acquired the refrigerant recovery or recycling device(s) listed in Part 2, that the establishment is complying with Section 608 regulations, and that the information given is true and correct.

Signature of Owner/Responsible Officer   Date   Name (Please Print)   Title

Instructions

Part 1: Please provide the name, address, and phone number of the establishment where the refrigerant recovery or recycling device(s) is (are) located. Please complete one form for each location. State the number of vehicles based at this location that are used to transport technicians and equipment to and from service sites.

Part 2: Check the appropriate boxes for the type of work performed by technicians who are employees of the establishment. The term "small appliances" refers to any of the following products that are fully manufactured, charged, and hermetically sealed in a factory with five pounds or less of refrigerant: refrigerators and freezers designed for home use, room air conditioners (including window air conditioners and packaged terminal air conditions), packaged terminal hear pumps, dehumidifiers, under-the-counter ice makers, vending machines, and drinking water coolers.

Part 3: For each recovery or recycling device, please list the name of the manufacturer of the device, and (if applicable) its model number and serial number.

If more than 7 devices have been acquired, please fill out an additional form and attach it to this one. Recovery devices that are self-contained should be listed first and should be identified by checking the box in the last column on the right. A self-contained device is one that uses its own pump or compressor to remove refrigerant from refrigeration or air conditioning equipment. On the other hand, system-dependent recovery devices rely solely upon the compressor in the refrigeration or air conditioning equipment and/or on upon the pressure of the refrigerant inside the equipment to remove the refrigerant inside the equipment to remove the refrigerant.

If the establishment has been listed as Type B and/or Type D in Part 2, then the first device listed in Part 3 must be a self-contained device and identified as such by checking the box in the last column to the right.

If any of the devices are homemade, they should be identified by writing "homemade" in the column provided for listing the name of the device manufacturer. Homemade devices can be certified for establishments that are listed as Type A or Type B in Part 2 until [six months after promulgation of the rule]. Type C or Type D establishments can certify homemade devices at any time. If, however, a Type C or Type D establishment is certifying equipment after [six months after promulgation of the rule], then it must not use these devices for service jobs classified as Type A or Type B.

Part 4: This form must be signed by either the owner of the establishment or another responsible officer. The person who signs is certifying that the establishment is complying with Section 608 regulations, and that the information provided is true and correct.

EPA Regional Offices

Send your form to the EPA office listed below under the state or territory in which the establishment is located:

Connecticut, Maine, Massachusetts, New Hampshire, Rhode Island, Vermont

CAA 608 Enforcement Contact: EPA Region I, Mail Code APC, JFK Federal Building, One Congress Street, Boston, MA 02203

New York, New Jersey, Puerto Rico, Virgin Islands

CAA 608 Enforcement Contact: EPA Region II, Jacob K. Javits Federal Building Room 5000, 26 Federal Plaza, New York, NY 10278

Delaware, District of Columbia, Maryland, Pennsylvania, Virginia, West Virginia

CAA 608 Enforcement Contact: EPA Region III, Mail Code 3AT21, 841 Chestnut Building, Philadelphia, PA 19107

Alabama, Florida, Georgia, Kentucky, Mississippi, North Carolina, South Carolina, Tennessee

CAA 608 Enforcement Contact: EPA Region IV, Mail Code APT-AE, 345 Courtland St. NE, Atlanta, GA 30365

Illinois, Indiana, Michigan, Minnesota, Ohio, Wisconsin

CAA 608 Enforcement Contact: EPA Region V, Mail Code AT18J, 77 W Jackson Blvd., Chicago, IL 60604

Arkansas, Louisiana, New Mexico, Oklahoma, Texas

CAA 608 Enforcement Contact: EPA Region VI, Mail Code 6T-EC, First Interstate Tower at Fountain Place, 1445 Ross Ave., Suite 1200, Dallas, TX 75202

Iowa, Kansas, Missouri, Nebraska

CAA 608 Enforcement Contact: EPA Region VII, Mail Code ARTX/ARBR, 726 Minnesota Ave., Kansas City, KS 66101

Colorado, Montana, North Dakota, South Dakota, Utah, Wyoming

CAA 608 Enforcement Contact: EPA Region VIII, Mail Code 8AT-AP, 999 18th Street, Suite 500, Denver, CO 80202

America Samoa, Arizona, California, Guam, Hawaii, Nevada

CAA 608 Enforcement Contact: EPA Region IX, Mail Code A-3, 75 Hawthorne St., San Francisco, CA 94105

Alaska, Idaho, Oregon, Washington

CAA 608 Enforcement Contact: EPA Region X, Mail Code AT-082, 1200 Sixth Ave., Seattle, WA 98101

Abrasion

A scrape or other damage on an object's surface.

Acronym

Letters that stand for a phrase. *HVACR stands for heating, ventilation, air conditioning, and refrigeration; CFC stands for chlorofluorocarbon.*

Alkylbenzene

An organic lubricant that's made from the raw chemicals propylene, a colorless hydrocarbon gas, and benzene, a colorless liquid hydrocarbon.

Ambient temperature

Temperature of the air around an object. *Ambient* comes from a Latin word that means "to surround."

Analogous

From the word *analogy*—a comparison of two different things that are alike in some way.

Analysis

Separating the parts of a thing to study them separately. *If someone analyzes iron ore they can find iron.*

Anticipate

To look forward to, or to expect. From a Latin word that means "to take before."

Antidote

A remedy that counteracts the effects of a poison.

Aperture

A service connection used to access a sealed refrigeration system, like a clamp-on piercing valve.

Appliance

A broad term used for electrical devices, including air-conditioning and refrigeration units—a refrigerator, freezer, central air conditioner, walk-in cooler, or centrifugal chiller.

Atmospheric pressure

The pressure caused by the weight of the air above a certain point. Normal atmospheric pressure at sea level is about 15 pounds per square inch.

Atom

The smallest unit of a chemical element. Every atom is made up of a positively charged *nucleus* and a set of negatively charged *electrons* that revolve around the nucleus. The nucleus is made up of positively charged *protons* and *neutrons* that have no charge. Atoms link together to form *molecules*.

Attrition

A natural rubbing away or wearing down. From a Latin word that means "to rub against."

Azeotrope

A constant-boiling mixture. A mixture of two liquids that boils at constant composition. The vapor's composition is the same as the liquid's. When the mixture boils, at first the vapor has a higher proportion of one component than is present in the liquid, so this proportion in the liquid falls over time. Eventually, maximum and minimum points are reached, at which the two liquids distill together with no change in composition. An azeotrope's composition depends on pressure.

Binary

Anything made up of two parts. From a Latin word meaning "two by two."

Boil

To change from a liquid to a vapor.

Calibrate

To systematically adjust the graduations of a measuring instrument.

Certified Refrigerant Recycling or Recovery Equipment

Equipment certified by an approved equipment testing organization to meet EPA standards.

Chlorine monoxide

A molecule made of one oxygen atom and one chlorine atom. It's found in the stratosphere when ozone depletion is taking place. By measuring chlorine monoxide, scientists can determine the degree of ozone depletion.

Chlorofluorocarbon (CFC)

Any of several compounds made up of carbon, chlorine, and fluorine. CFCs were used as aerosol propellants and refrigerants until they were found to be harmful to the earth's protective ozone layer.

Commercial Refrigeration

Refrigeration equipment utilized in the retail food and cold storage warehouse sectors.

Compatible

Capable of orderly, efficient integration and operation with other elements in a system.

Comply

To act in accord with a rule, standard, or law.

Compound

In chemistry, a substance that contains two or more elements in definite proportions. Only one molecule present in a compound.

Condense

To change from a vapor to a liquid. From a Latin word that means "to thicken."

Configuration

An arrangement of elements or parts in a system.

Constituent

One part of a whole; a component.

Constituents

Parts of a whole.

Contaminants

Dirt, moisture, or any other substance that is foreign to a refrigerant.

Contingent

Depends on conditions or events that may or may not happen; conditional.

Corresponding

Agreeing or conforming, as in degree or kind.

Critical temperature

The highest temperature a gas can have and still be condensable by pressure.

Deficient

Lacking an essential element; incomplete.

DeMinimus

Minimum. The smallest quantity, number, or degree possible or permissible.

Dielectric

A material that conducts electricity little or not at all. If a voltage is applied to a dielectric, atoms in the material arrange themselves to oppose the flow of electric current. Glass, wood, and plastic are common dielectrics. *Dielectric strength* is a measure of resistance that materials such as oil have to electric current.

Dispose

To get rid of something.

Disposable cylinder

A one-trip refrigerant cylinder; not to be refilled.

Distill

The process of separating constituents of a liquid by boiling it then condensing the vapor that's produced. Distillation can be used to purify water and other substances, or to remove one component from a mixture, as when gas is distilled from crude oil.

Ells

Pre-bent tubing, factory-designed as soldered or flared fittings that let refrigerant lines be routed between components.

Entity

Something that exists as a single unit. Refrigerant owned by an entity would be the property of one person or corporation.

Equalizer valve

A device that regulates the flow of gases or liquids. It's used to balance pressures on either side of some recovering machines.

Ester Oil

An oil used with hydrofluorocarbon (HFC) refrigerants.

Evacuate

To remove air (gas) and moisture from a refrigeration or air conditioning system.

Excerpt

A selected passage from an article or book.

Excise tax

An internal tax such as the tax levied on those who produce chlorofluorocarbons (CFCs).

Expansion valve

A metering device used in refrigeration and air-conditioning applications that separates the system's low and high sides.

Five-carbon neopentyl alcohols

An alcohol compound with a five-carbon molecular structure.

Floc point

The temperature at which wax separates out (precipitates) from a mixture of 10 percent oil and 90 percent refrigerant. The floc point is a measure of an oil's relative tendency to separate wax when mixed with a soil-soluble refrigerant.

Fluorocarbon

A molecule that contains fluorine and carbon atoms.

Fractionation

When one or more refrigerants of the same blend leak at a faster rate than other refrigerants in the blend, changing the composition of the blend. *Fractionation is possible only when liquid and vapor exist at the same time.*

Frangible disk

A circular (round and dished) device used on some refrigeration equipment to provide pressure release for safety purposes. The frangible disk suddenly breaks when a certain pressure is reached.

Fully halogenated CFC

When all the hydrogen atoms in a hydrocarbon molecule are replaced with chlorine or fluorine atoms.

Girth steam

A welded or bolted connection between a tube's lower and upper halves, placed near the center or widest diameter.

Global warming

Often called the *greenhouse effect*. In global warming, tropospheric pollutants like CFCs, HCFCs, HFCs, carbon dioxide, and carbon monoxide, absorb and reflect the earth's infrared radiation. This causes radiation back to the earth, and a gradual increase in the earth's average temperature.

Halogen

Any of the five chemically related nonmetallic elements that include fluorine, chlorine, bromine, iodine, and astatine.

Halogenate

To cause some other element to combine with a halogen.

Hermetic

Totally sealed, especially against the escape or entry of air. In HVACR applications, it means sealed by gaskets or welds, as in refrigeration compressors.

Hermetically sealed

Any object or substance confined in a gas- or air-tight container. A refrigeration system is hermetically sealed.

High-pressure appliance

An appliance that uses a refrigerant with a boiling point between $-50°C$ and $10°C$ at atmospheric pressure.

Hydrocarbon

A molecule that contains hydrogen and carbon atoms. An organic compound containing only hydrogen and carbon.

Hydrochlorofluorocarbons (HCFCs)

Molecules created when some of the hydrogen atoms in a hydrocarbon molecule are replaced with chlorine or fluorine atoms. Because they have a shorter life than CFCs, HCFCs are less harmful the CFCs to stratospheric ozone.

Hydrodynamic lubrication

Lubrication that deals with the motion of fluids and the behavior of solid bodies immersed in them.

Hydrofluorocarbons (HFCs)

Molecules created when some of the hydrogen atoms in a hydrocarbon are replaced with fluorine. Because HFCs contain no chlorine, they don't destroy ozone but they contribute to global warming.

Hydrostatically tested

A process used to test the bursting points of cylinders or tanks (pressure vessels). They're filled with fluid, tightly closed, the subjected to a calibrated pressure.

Hygroscopic

Readily absorbs and retains moisture, as from the atmosphere.

Incompatible

Not suited to be used together; not in harmony or agreement.

Incorporate

To thoroughly blend or combine into an existing thing. From a Latin word that means "to form into a body."

Industrial-process refrigeration

Complex, customized appliances used in the chemical, pharmaceutical, petrochemical, and manufacturing industries.

Inert

An inert chemical is one that shows no chemical activity except under extreme conditions. For example, CFCs are inert; they have a long life and are broken up into chlorine, fluorine, and carbon atoms only when they're exposed to ultraviolet light in the stratosphere.

Interim

Short term or temporary. From a Latin word for "in the meantime."

Isomers

Molecules that have the same numbers of the same atoms, but the atoms are arranged differently in their structure. Even though isomers of the same compound have equal numbers of atoms of the same element, they have very different physical properties.

King valve

A liquid-receiver outlet service valve.

Latent heat

Heat that can't be measured with a thermometer. Latent heat is hidden heat that's generated when substances change states.

Low-loss fittings

Any device that connects hoses, appliances, or recovery or recycling machines, and that is designed to close automatically or to be closed manually when it's disconnected.

Low pressure appliance

An appliance that uses a refrigerant with a boiling point above 50°F at atmospheric pressure.

Malignancy

Abnormal mass of new tissue growth that serves no function in the body and that threatens life or health.

Miscible

Capable of being mixed in all proportions.

Mixture

A blend of two or more components that don't have a fixed proportion to each other and that, however well blended, keep their individual chemical characteristics. Unlike compounds, mixtures can be separated by physical methods like distillation. *One example is a near-azeotropic blend of refrigerants.*

Moisture indicator

An instrument used to measure a refrigerant's moisture content.

Molecule

A stable configuration of atoms held together by electrostatic and electromagnetic forces. A molecule is the simplest structural unit that displays a compound's characteristic physical and chemical properties.

Montreal Protocol

An agreement signed in 1987 by the United States and 22 other countries, and updated several times since then, to control releases of ozone-depleting substances like CFCs and HCFCs, and eventually phase out their use.

Near-Azeotropic blend

A blend that acts very much like an azeotrope, but has a small volumetric composition change and temperature glide as it evaporates and condenses.

Nomenclature

A system of special terms or symbols, like those used in science. Nomenclature comes from the Latin word "nomenclator," a slave who accompanied his master to tell him the names of people he met.

Non-condensable gas

Gas that doesn't change to a liquid at operating temperatures and pressure.

Non-miscible

When two substances, such as oil and water, are incapable of mixing.

Non-polar

A system or substance without opposite extremes, as of magnetism or electric charge.

Obsolete

No longer in use; outmoded in style, design, or construction.

Organic

Something derived from living organisms.

Oxidation

Any chemical reaction where a substance gives up electrons—as when a substance combines with oxygen. *Burning is an example of fast oxidation; rusting is an example of slow oxidation.*

Oxidize

A corrosive chemical reaction caused by exposure to oxygen gas; like rust (iron oxide) or copper oxide (which forms on or inside copper tubing).

Ozone depletion

Happens when ultraviolet radition in the stratosphere breaks CFC and HCFC refrigerants into their atomic elements—chlorine, fluorine, and hydrogen atoms. Chlorine atoms react with and destroy stratospheric ozone, which protects earth's human and other life forms from the sun's harmful ultraviolet radition.

Partially halogenated HCFC

When not every hydrogen atom in a hydrogen molecule is replaced with chlorine or fluorine atoms.

Particle

A tiny piece or part; speck.

Permeability

An object's or substance's ability to be penetrated.

Phenomenon

An event or fact directly perceptible by the senses.

Photochemical reaction

A chemical reaction caused by light or ultraviolet radiation.

Phytoplankton and larvae

Plants, organisms, and newly hatched insects that float on the ocean's surface and are a source of food for fish and other marine life.

Placard

An easily seen tag or label that usually indicates warning or caution.

Polar

Molecules that have a positively charged end and a negatively charged end, each of which attracts its opposite.

Polyalkylene glycols (PAGs)

A very hygroscopic refrigeration lubricant for use with HFC refrigerants. Used often in automotive air conditioning systems when employing HFC refrigerants. PAGs are incompatible with chlorine and have very high molecular weights.

Polymer

A long molecule made up of a chain of smaller, simpler molecules.

Polyolesters

Polyolesters have stable five-carbon neopentyl alcohols which, when mixed with fatty acids, will form the polyol ester family. A popular synthetic lubricant for use with HFC refrigerants. Used as a jet engine lubricant for years.

Pressure vessel

A holding device that maintains a certain "force per unit of area."

Process stub

A tube that extends from the compressor or filter drier of a hermetic system. It's used to gain access to the sealed system.

Proprietary

Sole ownership of property, a business, an item of labor, or an object that extends legal ownership rights. From Latin words that mean "property" and "one's own."

Quick-disconnect fittings

Fittings used on refrigerant hoses that seal automatically when removed from an appliance. *Quick-disconnect fittings will help reduce refrigerant losses when removing hoses.*

Ream

To scrape, cut, or otherwise clean the inside of pipe or copper tubing. Reaming eliminates ridges or raised surfaces that come from cutting the pipe or tubing.

Reclamation

Restoring to usefulness.

Refrigerant receiver

A refrigeration system component installed in the liquid line. It's designed to make a space for liquid refrigerant flow due to the closing action of a self-regulating metering device.

Regulatory authority

The right, as for a government agency, to control an industrial process or mechanism in agreement with a rule.

Residue

A substance left over at the end of a process. For example, *residual oil* is the low-grade oil product left after gasoline is distilled.

Retrofit

To furnish with new equipment or parts that weren't available when a device or system was first manufactured.

Revoke

To cancel or take back a permission or contract. From a Latin word that means "to call back."

Saturated pressure

The force in a pressure vessel that matches the temperature of a certain contained gas at a condition where any removed heat would cause condensation, and added heat would cause evaporation.

Saturation temperature

The temperature at which a liquid turns to vapor or a vapor turns to a liquid.

Schrader valve

Valves that use a valve core, like a tire-valve stem, to gain access to a sealed system. *Schrader valves help HVACR technicians recover refrigerant.*

Silkscreen

A stencil-producing method where a design is imposed on a screen of silk. Blank areas are coated with an impenetrable substance, and ink is forced through the cloth onto the printing surface.

Soluble

A substance that can be dissolved in a given liquid.

Specific density

In reference to a refrigerant, a statement of mass per unit of volume measurement (pounds per cubic foot).

Specific volume

In reference to refrigerant, a corresponding but reversed value of specific density (cubic feet per pound).

Spring-loaded relief

A calibrated spring-operated pressure release device that's designed to relieve pressure for safety purposes.

Stratosphere

The atmosphere between 7 and 30 miles above the earth where a layer of ozone filters out harmful ultraviolet light.

Subcooling

A liquid below its saturation temperature for a certain saturation pressure.

Superheat

A vapor above its saturation temperature for a certain saturation pressure.

Symmetry

Having balanced, evenly distributed parts.

Synthetic

Produced artificially. In chemistry, forming a compound from its parts.

System-Dependent Recovery Equipment

Refrigerant recovery equipment that requires the assistance of components contained in an appliance to remove the refrigerant from the appliance.

Technical bulletins

Published industrial statements about newsworthy items.

Technician

Any person who performs maintenance, service, or repair that could reasonably be expected to release Class I or Class II substances from appliances into the atmosphere, including but not limited to installers, contractor employees, in-house service personnel, and in some cases, owners.

Temperature glide

Range of condensing or evaporating temperatures for one pressure.

Terminate

To bring to an end.

Ternary

Having three elements, parts, or divisions.

Total Equivalent Warning Impact (TEWI)

A unit of measurement that assesses the total effect CFCs, HCFCs, and HFCs have on global warming.

Thermodynamics

The physics of the relationship between heat and other forms of energy.

Throttle

To slowly obstruct flow.

Toxicology

The study of poisons, their effects, and antidotes.

Transition

The process of changing from one state or form to another.

Troposphere

The lowest level of the atmosphere—from the ground to seven miles above the earth—where ultraviolet rays from the sun react with pollution and smog to form ozone.

Ultraviolet radiation

Radiation in the part of the electromagnetic spectrum where wavelengths are shorter than visible violet light but longer than X-rays. UV radiation causes cancer.

Vapor pressure

Pressure applied to a saturated liquid.

Vehicle

A device—like a car or truck—for moving passengers and all kinds of goods and products from one place to another.

Verify

To prove accuracy by observing, testing, and presenting evidence.

Very high pressure appliance

An appliance which uses a refrigerant with a boiling point below -50°C at atmospheric pressure.

Virgin refrigerant

New, original, non-recycled refrigerant.

Volatile

Vaporizes easily. A liquid's flash point—the temperature and pressure at which a liquid turns to gas.

Zeotrope

Refrigerant blends that change volumetric composition and saturation temperatures as they evaporate or condense at constant pressures. Zeotropes have a temperature glide as they evaporate and condense. (Zeotrope and non-azeotrope are synonyms.)

INDEX

condensing, 1

discharge, 1

evaporating, 1-3

gauge, 1-3

head, 1. (See also Condensing pressure)

high side, 1. (See also Condensing pressure)

low side, 1. (See also Evaporating pressure)

suction, 1. (See also Evaporating pressure)

vapor, 3-4

Process stub, 119

Program for Alternative Fluorocarbon Toxicity Testing, 134

P-traps, 9

Purge

loss from recycling equipment, 105

systems, 130-31

venting, 137

R-11 vapor recovery, 132-33

R-123, 39, 134, 139

R-124, 134

R-12 procedure for retrofit, 140

R-134a, 38-39, 134, 139-40

R-22 service equipment, 37

R-23, 38

R-401A, 39-41, 43-44, 71

R-401B, 39-40, 71

R-402A, 38, 40, 71

R-402B, 38, 40, 71

R-404A, 37-38, 71

R-406A, 39

R-407C, 37, 39, 71

R-407D, 38

R-408A, 38, 71

R-409A, 39

R-410A, 37-39, 71

R-414A, 39

R-414B, 39

R-416A, 39

R-502, 40

R-507, 38

R-508A, 38

R-508B, 38

Radiation

infrared, 68

ultraviolet, 60-64

Receiver, 5, 8

above condenser, 16, 29

Reclaim, 93

Record keeping, 82

Recovery, 93

active, 85-86, 91

cylinders

changing, 100-1

used in, 108

efficiency, 102, 104

equipment

safety, 104

used for, 102-3

levels, 89-90

machine

grandfathering provisions of, 105

standards, 104-6

passive, 86

requirements, 89-90

reverse, 95

speeding up the process, 86

typical, 96-97, 99

vacuum levels, 104

when not to, 89

Recycling, 93, 101-3

efficiency, 102, 104

equipment safety, 104

machine

grandfathering provisions of, 105

standards, 104-6

purge loss, 015

vacuum levels, 104

Refillable cylinders, 113

Refilling precautions, 115-16

Refrigerant, 33-47

blend nomenclature, 47

changing types and filters, 99-100

charging, 133

classification, 137-38

compatibility, 133

conditions, 3

containment options, 94

cylinders

disposable, 107-8

refillable, 113

returnable, 107-8

Temperature glide, 42
Ternary blends, 37, 40
TEV. (See Thermostatic expansion valve)
TEWI. (See Total equivalent warming impact)
Thermostat, 8
Thermostatic expansion valve, 8, 12-13
Three-way service valves, 14-15
Threshold Limit Value, 135
Time Weighted Average, 134
Title VII, 82-83
TLV. (See Threshold Limit Value)
Ton tank
    filling procedure, 111
    labeling and marking, 110-11
    retesting, 109
Total equivalent warming impact, 68-69
Toxicity characteristic, 56
Transportation
    of cylinders, 112
    vehicular, 117
Tropospheric
    ozone, 59-60
    pollutants, 65
TW. (See Tare weight)
TWA. (See Time Weighted Average)
TXV. (See Thermostatic expansion valve)

Ultrasonic leak detector, 123, 125-26
Ultraviolet radiation. (See also Radiation)
    types, 60
UNEP. (See United Nations Environment Programme)
United Nations Environment Programme, 64
Universal certification, 80
Use log, virgin refrigerant, 83
User information, cylinders, 111

Vacuum
    pumps, 88-89, 120-21
    valve and fitting, 104

Valve
    bypass solenoid, 8
    condenser water regulating 16, 23
    equalization, 104
    four-way reversing, 16, 24-26
    holdback, 16, 22
    inlet, and fitting, 104
    outlet, and fitting, 104
    Schrader, 15, 87, 119
    three-way service, 14-15
    vacuum, and fitting, 104
Vapor
    compression refrigeration system, 1-2
    point, saturated, 11
    pressure, 3-4
    pump, 11
    recovery, 95, 132-33
Vaporizing, 1. (See also Evaporation)
Vehicular transportation of cylinders, 117
Ventilation, 137
Venting
    prohibition on, 119
    purge, 137
    report, accidental or unintentional, 83-84
Very high pressure appliance (Type II), 80
Vibration eliminator, 121
Viscosity, oil, 50-51
Volume-sensitive shutoff, 105

Warming, global, 65-69
Warning labels, 112-13
Waste oil, 56-57
Water capacity weight, 114-15
WC. (See Water capacity weight)
Welding, 136-37

Zeotropic mixtures, 134